ASIMOV ON ASTRONOMY

In some ways, science fiction writers aren't doing too well these days. In late 1962, Mariner II seemed to settle the question of the surface temperature of Venus placing it far above the boiling point of water.

With that, there vanished some of the most beautiful settings for s.f. stories. Old-timers may remember with nostalgia, as I do, the moist swampy world of Weinbaum's 'Parasite Planet'. Well it's gone! For that matter, I wrote a short novel under a pseudonym, a number of years ago, that was set on a Venus that was one huge ocean, with Earth-cities built underwater in the shallower regions ... All gone!

Now along comes Mariner IV and discovers craters (but no canals) on Mars.

No one expected that! I don't know of a single science fiction story that had ever placed craters on Mars ... Canals, yes, but craters, no! I have written several stories set on Mars and I have always mentioned the canals (I placed no water in them; I knew enough for that) but I've never had craters.

**Also by the same author,
and available in Coronet Books:**

The Tragedy of the Moon

Asimov on Astronomy

Isaac Asimov

CORONET BOOKS
Hodder and Stoughton

First published in Great Britain 1974 by
Macdonald and Jane's, Macdonald and
Company (Publishers) Ltd

Coronet Edition 1976

Printed and bound in Great Britain for
Coronet Books, Hodder and Stoughton, London
By Cox & Wyman Ltd, London, Reading
and Fakenham

ISBN 0 340 20015 4

To Paul R. Esserman,
easily the best internist in the world

ILLUSTRATIONS

Acknowledgements are made to the following:

Craters on the Moon; *National Aeronautics and Space Administration.*

Man on the Moon; *National Aeronautics and Space Administration.*

Isaac Newton; *The Bettmann Archive.*

Edmund Halley; *The Bettmann Archive.*

Saturn; *Mount Wilson and Palomar Observatories.*

Jupiter; *Mount Wilson and Palomar Observatories.*

Venus; *Mount Wilson and Palomar Observatories.*

Pluto; *Mount Wilson and Palomar Observatories.*

Star Cluster; *Mount Wilson and Palomar Observatories.*

Galactic Clusters; *Mount Wilson and Palomar Observatories.*

The Crab Nebula; *Hale Observatories.*

Spiral Galaxies; *Hale Observatories.*

Supernovae; *Lick Observatory.*

All essays in this volume are reprinted from *The Magazine of Fantasy and Science Fiction*. Individual essays appeared in the following issues:

CONTENTS

INTRODUCTION

Back in 1959, I began writing a monthly science column for *The Magazine of Fantasy and Science Fiction*. I was given *carte blanche* as to subject matter, approach, style, and everything else, and I made full use of that. I have used the column to range through every science in an informal and very personal way so that of all the writing I do (and I do a great deal) nothing gives me so much pleasure as these monthly essays.

And as though that were not pleasure enough in itself, why, every time I complete seventeen essays, Doubleday & Company Inc., puts them into a book and publishes them. As of this moment, I have had published nine books of my F & SF essays, containing a total of 153 essays. A tenth is, of course, in the works.

Few books, however, can be expected to sell indefinitely; at least not well enough to be worth the investment of keeping them forever in print. The estimable gentlemen at Doubleday have, therefore (with some reluctance, for they are fond of me and know how my lower lip tends to tremble on these occasions), allowed the first five of my books of essays to go out of print.

Out of *hardback* print, I hasten to say. All five of the books are flourishing in paperback so that they are still available to the public. Nevertheless, there is a cachet about the hardback that I am reluctant to lose. It is the hardbacks that supply the libraries; and for those who really want a permanent addition to their large personal collections of Asimov books[1] there is nothing like a hardback.

My first impulse, then, was to ask the kind people at Doubleday to put the books back into print and gamble on a kind of second wind. This is done periodically in the case of my science fiction books with success (even when paperback editions are simultaneously available). But I could see that the case was different. My science fiction is ever fresh, but science essays do tend to get out of date, for the advance of science is inexorable.

And then I got to thinking . . .

I deliberately range widely over the various sciences both to satisfy my own restless interests and to give each member of my heterogeneous audience a chance to satisfy his own particular taste now and then. The result is that each collection of essays has some on astronomy, some on chemistry, some on physics, some on biology, and so on.

But what about the reader who is interested in science, but is *particularly* interested in astonomy? He has to read through the non-astronomical articles in each book and can find only four or five, perhaps, on his favourite subject.

Why not, then, go through the five out-of-print books, cull out the astronomy articles, and put seventeen of them together in a volume we can call *Asimov on Astronomy*? Each individual article is old, but put together like that, the combination is new.

So here is the volume. It has six articles from *Fact and Fancy*, three from *View from a Height*, one from *Adding a Dimension*, four from *Of Time and Space and Other Things*, and three from *From Earth to Heaven*. The articles are arranged, not chronologically, but conceptually. As you turn the pages, you will find them taking on broader and broader scope. The first couple deal with Earth and its vicinity, the last couple with the Universe as a whole.

Aside from grouping the articles into a more homogeneous mass, what more have I done? Well, the articles are anywhere from six to thirteen years old and their age shows here and there. I feel rather pleased that the advance of science has not knocked out a single one of the articles here included, or even seriously dented any, but minor changes must be made, and I have made them.

In doing this I have not revised the articles themselves since that would deprive you of the fun of seeing me eat my words now and then, or, anyway, chew them a little. So I have made the changes by adding footnotes here and there where something I said needed modification or where I was forced to make a change to avoid presenting misinformation in the tables.

In addition to that, my good friends at Doubleday decided to prepare the book in more elaborate format than they have used for my ordinary essay collections, and have added illustrations to which I have written captions that give information above and beyond what is in the essays themselves.

Finally, since the subject matter is so much more homogeneous than in my ordinary grab-bag essay collections, I have prepared an index which will, I hope, increase the usefulness of the book as reference.

So, although the individual essays are old, I hope you find the book new and useful just the same. And at least I have explained, in all honesty, exactly what I have done and why. The rest is up to you.

ISAAC ASIMOV

New York, September 1972

CHAPTER ONE

TIME AND TIDE

What with one thing and another, I've got used to explaining various subtle puzzles that arise in connection with the scientific view of the universe. For instance, I have disposed of the manner in which electrons and photons can be waves part of the time and particles the rest of the time in a dozen different ways and by use of a dozen different analogies.

I've got so good at it, in fact, that at dinner parties the word nervously goes about, 'For heaven's sake, don't ask Asimov anything about wave-particle duality.'

And no one ever does. I sit there all primed and aching to explain, and no one ever asks. It kills the party for me.

But it's the simple thing that throws me. I've just been trying to write a very small book on the Moon[1] for third-graders and as part of the task I was asked to explain why there are two high tides each day.

Simple, I thought, and a condescending smirk passed over my face. I flexed my fingers and bent over the typewriter.

As the time passed, the smirk vanished and the hair at my temples grew perceptibly greyer. I managed at last, after a fashion, but if you don't mind, Gentle Reader, I'd like to try again. I need the practice.

The tides have bothered people for a long time, but not the good old Greeks, with reference to whom I start so many articles. The Greeks, you see, lived (and still live, for that matter) on the shores of the Mediterranean Sea. That sea happens to be relatively tideless because it is so nearly land-locked that high tide can't get through the Strait of Gibraltar before the time for it has passed and it is low tide again.

About 325 B.C., however, a Greek explorer, Pytheas of Massalia (the modern Marseilles), ventured out of the Mediterranean and into the Atlantic. There he came across good pronounced tides, with two periods of high water each day and two periods of low water in between. Pytheas made good observations of these, undoubtedly helped out by the inhabitants of the shores facing the open ocean who were used to the tides and took them for granted.

The key observation was that the range between high water and low water was not always the same. It increased and decreased with time. Each month there were two periods of particularly large range between high and low tides ('spring tides') and, in between, two periods of particularly small range ('neap tides').

What's more, the monthly variations matched the phases of the Moon. The spring tides came at full Moon and new Moon, while the neap tides came at first quarter and third quarter. Pytheas suggested, therefore, that the tides were caused by the Moon. Some of the later Greek astronomers accepted this, but for the most part, Pytheas's suggestion lay fallow for two thousand years.

There were plenty of men who believed that the Moon influenced the manner in which crops grew, the rationality or irrationality of men, the way in which a man might turn into a werewolf, the likelihood of encountering spooks and goblins — but that it might influence the tides seemed to be going a bit far!

I suspect that one factor that spoiled the Moon/tide

connection for thoughtful scholars was precisely the fact that there were two tides a day.

For instance, suppose there is a high tide when the Moon is high in the sky. That would make sense. The Moon might well be drawing the water to itself by some mysterious force. No one in ancient and medieval times had any notion of just how such a force might behave, but one could at least give it a name such as 'sympathetic attraction'. If the water heaped up under a high Moon, a point on the rotating Earth, passing through the heap, would experience a high tide followed by a low tide.

But a little over twelve hours later, there would be another high tide and then the Moon would be nowhere in the sky. It would be, in fact, on the other side of the globe, in the direction of a man's feet. If the Moon were exerting a sympathetic attraction, the water on the man's side of the globe ought to be pulled downwards in the direction of his feet. There ought to be a hollow in the ocean, not a heap.

Or could it be that the Moon exerted a sympathetic attraction on the side of the Earth nearest itself and a sympathetic repulsion on the side opposite. Then there would be a heap on both sides, two heaps all told. In one rotation of the Earth, a point on the shore would pass through both heaps and there would be two high tides each day, with two low tides in between.

The notion that the Moon would pull in some places and push in other places must have been very hard to accept, and most scholars didn't try. So the Moon's influence on the tides was put down to astrological superstition by the astronomers of early modern times.

In the early 1600s, for instance, Johannes Kepler stated his belief that the Moon influenced the tides, and the sober Galileo laughed at him. Kepler, after all, was an astrologer who believed in the influence of the Moon and the planets on all sorts of earthly phenomena and Galileo would have none of that. Galileo thought the tides were caused by the sloshing of the oceans back and forth as the Earth rotated — and he was quite wrong.

Came Isaac Newton at last! In 1685, he advanced the law of universal gravitation. By using that law it became obvious that the Moon's gravitational field had to exert an influence on the Earth and the tides could well be a response to that field.

But why *two* tides? What difference does it make whether we call the force exerted by the Moon on the Earth 'sympathetic attraction' or 'gravitational attraction'? How could the Moon, when it was on the other side of the Earth, cause the water on this side to heap upwards, *away* from the Moon. The Moon would still have to be pulling in one place and pushing in another, wouldn't it? And that still wouldn't make sense, would it?

Ah, but Newton did more than change words and substitute 'gravity' for 'sympathy'. Newton showed exactly how the gravitational force varied with distance, which was more than anyone before him had shown in connection with any vaguely postulated sympathetic force.

The gravitational force varied inversely as the square of the distance. That means the force grows smaller as the distance grows larger; and if the distance increases by a ratio of x, the force decreases by a ratio of x^2.

Let's take the specific case of the Moon and the Earth. The average distance of the Moon's centre from the surface of the Earth nearest itself is 234,000 miles. In order to get the distance of the Moon's centre from the surface of the Earth farthest from itself, you must add the thickness of the Earth (8,000 miles) to the first figure, and that gives you 242,000 miles.

If we set the distance of the Moon to the near surface of

the Earth at 1, then the distance to the far surface is 242,000/234,000 or 1·034. As the distance increases from 1·000 to 1·034, the gravitational force decreases from 1·000 to $1/1·034^2$, or 0·93.

There is thus a 7·0 per cent difference in the amount of gravitational force exerted by the Moon on the two sides of the Earth.

If the Earth were made of soft rubber, you might picture it as yielding somewhat to the Moon's pull, but each part would yield by a different amount depending on the strength of the pull on that particular part.

The surface of the Earth on the Moon's side would yield most since it would be most strongly attracted. The parts beneath the surface would be attracted with a progressively weaker force and move less and less towards the Moon. The opposite side of the Earth, being farthest from the Moon would move towards it least of all.

There would therefore be two bulges; one on the part of the Earth's surface nearest the Moon, since that part of the surface would move the most; and another on the part of the Earth's surface farthest from the Moon, since that part of the surface would move the least and lag behind all the rest of the Earth.

If that's not clear, let's try analogy. Imagine a compact group of runners running a long race. All of them run towards the finish line so that we might suppose some 'force' is attracting them towards that finish line. As they run, the speedier ones pull out ahead and the slower ones fall behind. Despite the fact that only one 'force' is involved, a 'force' directed towards the finish line, there are two 'bulges' produced; a bulge of runners extending forward towards the finish line in the direction of the force, and another bulge of runners extending backwards in the direction opposite to that of the force.

Actually the solid body of the Earth, held together by strong intermolecular forces, yields only very slightly to the gravitational differential exerted by the Moon on the Earth. The liquid oceans, held together by far weaker inter-molecular forces, yield considerably more and make two 'tidal bulges', one towards the Moon and one away from it.

As the Earth rotates, an individual point on some seacoast is carried past the first tidal bulge and then half a day later through the second. There are thus two high tides and two low tides in one complete rotation of the Earth — or, to put it more simply, in one day.

If the Moon were motionless, the tidal bulges would always remain in exactly the same place, and high tides would be exactly twelve hours apart. The Moon moves in its orbit about the Earth, however, in the same direction that the Earth rotates, and the tidal bulges move with it. By the time some point on Earth has passed through one bulge and is approaching a second, that second bulge has moved onward so that the Earth must rotate an additional half hour in order to pass the point under question through high tide again.

The time between high tides is twelve hours and twenty-five minutes, and the time from one high tide to the next but one is twenty-four hours and fifty minutes. Thus, the high tides each day come nearly one hour later than on the day before.

But why spring tides and neap tides and what is the connection between tides and the phases of the Moon?

For that we have to bring in the Sun. It, too, exerts a gravitational influence on the Earth. The gravitational pull of two separate heavenly bodies on the Earth varies directly with the mass of the bodies in question and inversely with the square of their distance from the Earth.

To make things simple, let's use the mass of the Moon as

the mass-unit, and the average distance of the Moon from the Earth (centre to centre) as the distance-unit. The Moon possesses 1 Moon-mass and is at 1 Moon-distance in other words, and the Moon's gravitational pull upon us can therefore be set at $1/1^2$ or 1.

The mass of the Sun is 27,000,000 times that of the Moon and its distance from the Earth is 392 times that of the Moon. We can say, then, that the Sun is 27,000,000 Moon-masses and is at 392 Moon-distances. The gravitational pull of the Sun upon the Earth is therefore $27,000,000/392^2$ or 176. This means that the Sun's gravitational pull upon the Earth is 176 times that of the Moon. You would therefore expect the Sun to create tidal bulges on the Earth, and so it does. One bulge on the side towards itself, naturally, and one on the side opposite itself.

At the new Moon, the Moon is on the same side of the Earth as the Sun, and both Moon and Sun are pulling in the same direction. The bulges they produce separately add to each other, producing an unusually large difference between high and low tide.

At the full Moon, the Moon is on the side of the Earth opposite that of the Sun. Both, however, are producing bulges on the side nearest them *and* on the side opposite them. The Sun's near-bulge coincides with the Moon's far-bulge and vice versa. Once again, the bulges produced separately add to each other and another unusually large difference between high and low tide is produced.

Therefore the spring tides come at new Moon and full Moon.

At first and third quarter, when the Moon has the half-Moon appearance, Moon, Earth, and Sun form a right triangle. If you picture the Sun as pulling from the right and producing a tidal bulge to the right and left of the Earth, then the Moon at first quarter is pulling from above and producing a bulge up and down. (At third quarter, it is pulling from below and still producing a bulge up and down.)

In either case, the two sets of bulges tend to neutralise each other. What would ordinarily be the Moon's low tide is partially filled by the existence of the Sun's high tide, so that the range in water level between high and low tide is cut down. Thus we have the neap tides at first and third quarter.

But hold on. I said that the Moon's low tide is 'partially filled' by the existence of the Sun's high tide. Only 'partially'. Does that mean the Sun's tidal bulges are smaller than the Moon's tidal bulges?

It sure does. The tides follow the Moon. The Sun modifies the Moon's effect but never abolishes it.

Surely, one ought to ask why that should be so. I have said that the Sun's gravitational pull on the Earth is 176 times that of the Moon. Why then should it be the Moon that produces the major tidal effect?

The answer is that it is not the gravitational pull itself that produces the tides, but the *difference* in that pull upon different parts of the Earth. The difference in gravitational pull over the Earth's width decreases rapidly as the body under consideration is moved farther off, since, as the total distance increases, the distance represented by the width of the Earth makes up a smaller and smaller part of the total.

Thus, the distance of the Sun's centre from the Earth's centre is about 92,900,000 miles. The Earth's width makes far less difference in this case than in the case, earlier cited, of the Moon's distance. The distance from the Sun's centre to the side of the Earth near it is 92,896,000, while the distance to the far side is 92,904,000. If the distance from the Sun's centre to the near side of the Earth is set equal to 1, then the distance to the far side is 1·00009. In that distance, the Sun's gravitational pull drops off to only $1/1 \cdot 00009^2$ or $0 \cdot 99982$.

In other words, where the difference in the Moon's gravitational pull from one side of the Earth to the other is $7 \cdot 0$ per cent; the difference of the Sun's gravitational pull is only $0 \cdot 018$ per cent. Multiply the Sun's gravitational difference by its greater gravitational pull overall ($0 \cdot 018 \times 176$) and you get $3 \cdot 2$ per cent. The tide-producing effect of the Moon is to that of the Sun as $7 \cdot 0$ is to $3 \cdot 2$ or as 1 is to $0 \cdot 46$.

We see then that the Moon's effect on tides is more than twice that of the Sun, despite the Sun's much greater gravitational pull.

A second way of attaining the comparative gravitational pulls of two bodies upon the Earth is to divide their respective masses by the *cubes* of their respective distances.

Thus, since the Moon has 1 Moon-mass and is at 1 Moon-distance, its tide-producing effect is $1/1^3$ or 1. The Sun with 27,000,000 Moon-masses at 392 Moon-distances, has a tide-producing effect of $27,000,000/392^3$ or $0 \cdot 46$.

We can easily see that no body other than the Sun and the Moon can have any significant tidal effect on the Earth. The nearest sizable body other than those two is the planet Venus. It can approach as closely as 26,000,000 miles, or 108 Moon-distances, at year-and-a-half intervals. Even then its tidal effect is only $66/109^3$ or $0 \cdot 0000051$ times that of the Moon.

The tides, in a way, affect time. At least, it is the tides that make our day twenty-four hours long. As the tidal bulge travels about the Earth, it scrapes against shallow sea bottoms (the Bering Sea and the Irish Sea are supposed to be the prime culprits) and the energy of Earth's rotation is dissipated as frictional heat. The energy of the Earth's rotation is so huge that this dissipation represents only a very small portion of the total over any particular year or even any particular century. Still, it is enough to be slowing the Earth's rotation and lengthening the day by one second every one hundred thousand years.

This isn't much on the human time scale, but if the Earth has been in existence for five billion years and this rate of day-lengthening has been constant throughout, the day has lengthened a total of fifty thousand seconds or nearly fourteen hours. When the Earth was created, it must have been rotating on its axis in only ten hours (or less, if the tides were more important in early geologic times than they are now, as they well might have been).

As the Earth's rate of rotation slows down, it loses angular momentum as well, but this angular momentum cannot be dissipated as heat. It must be retained, as angular momentum, elsewhere in the Moon-Earth system. What the Earth loses the Moon must gain and it can do this by receding from the Earth. Its greater distance means a greater angular momentum as it turns, since angular momentum depends not only upon rate of turn, but also upon distance from the centre about which an object is turning.

The effect of the tides, then, is to slow the Earth's rotation and to increase the distance of the Moon.

There is a limit to how much the Earth's rotation will be slowed. Eventually, the Earth will rotate about its axis so slowly that one side will always face the Moon as the Moon turns in its orbit. When that happens, the tidal bulges will be 'frozen' into place immediately under the Moon (and on Earth's opposite side) and will no longer travel about the Earth. No more friction, no more slowing. The length of the Earth day will then be more than fifty times as long as the present day; and the more distant Moon will turn in its orbit in twice the period it now turns.

Of course, the tidal bulges of the Sun will still be moving

about the Earth some seven times a year and this will have further effects on the Earth-Moon system, but never mind that now.

Even if there were no oceans on the Earth, there would still be tidal friction, for the solid substance of the Earth does yield a bit to the differential pull of the Moon. This bulge of solid material travelling around the Earth also contributes to internal friction and to the slowing of the Earth's rotation.

We can see this at work on the Moon, which has no oceans. Just as the Moon produces tides on the Earth, so the Earth produces tides on the Moon. Since the mass of the Earth is eighty-one times that of the Moon, but the distance is the same one way as the other, you might suspect that the tidal effect of Earth-on-Moon would be eighty-one times that of Moon-on-Earth. Actually, it's not quite that high. The Moon is a smaller body than the Earth so there's a smaller gravitational difference over its width than there would be in the case of the larger Earth. Without going into the details of the mathematics (after all, I must spare you something) I can give you the results —

If the effect of Moon-on-Earth is considered to be 1·00, then the effect of Earth-on-Moon is 32·5.

With the Moon affected 32·5 times as much as the Earth is, and with its mass, and therefore its rotational energy, considerably less than that of the Earth, there has been ample time in the history of the solar system to dissipate its rotational energy to the point where the tidal bulge is frozen into the Moon, and where the Moon faces one side only towards the Earth. This is actually the situation.

We can suspect that any satellite which receives a tidal effect even greater than that received by the Moon would (unless it were very much larger than the Moon) also face one side to its primary at all times.

As a matter of fact, there are six other satellites in the solar system that are Moon-sized or a little larger, and each of them is attached to a planet considerably more massive than the Earth. They are therefore much more affected tidally. If we continue to consider the effect of the Moon on the Earth to be 1·00, we have Table 1:

Table 1

Neptune-on-Triton	720
Saturn-on-Titan	225
Jupiter-on-Callisto	225
Jupiter-on-Ganymede	945
Jupiter on Europa	145
Jupiter on Io	5,650

There seems no question but that all these satellites have had their rotations with respect to their primaries stopped. Each of these satellites faces one side to its primary constantly.

What about the reverse, though? What about the effect of the various satellites on their primaries?

Of the six Moon-sized satellites just mentioned, the two which are closest to their primaries are Io and Triton. Io is 262,000 miles from Jupiter and Triton is 219,000 miles from Neptune. Because the effect varies inversely with the cube of the distance, we can suspect that these two will have considerably more effect on their primaries than will the remaining four, which are all much farther away from their primaries.

If we consider the Jupiter/Io and Neptune/Triton pairs, then we note that Jupiter is far larger than Neptune and that there will therefore be a larger drop in the gravitational field across the width of Jupiter than across the lesser width of Neptune. Since the extent of this drop is crucial, it is fair to conclude that of the six planet/satellite combinations we have been considering, the tidal effect of Io on Jupiter is the strongest. Let's see how much that is.

Again, we are considering the tidal effect of the Moon on the Earth to be 1·00. In that case (if you will trust my calculations) the effect of Io on Jupiter is equal to 30.

This is a sizable amount, surprising to anyone who would assume, without analysis, that a small satellite like Io could scarcely have much of a gravitational effect on giant Jupiter.

Well, it has. It has thirty times the effect on Jupiter that our Moon has on the Earth. Io exerts roughly the effect on Jupiter that the Earth exerts on the Moon.

Naturally, although the Earth's effect is sufficient to stop the Moon's rotation relative to itself, we wouldn't expect Io's similar effect on Jupiter to slow Jupiter's rotation significantly. After all, Jupiter is far larger in mass than the Moon is and packs far more rotational energy into its structure. Jupiter can dissipate this rotational energy for billions of years without slowing its rotation much while the Moon, dissipating its rotational energy at the same rate is brought to a halt. And, indeed, Jupiter still rotates with a period of only ten hours.

However, there are tidal effects other than rotation-slowing. It has recently been discovered that Jupiter's emission of radio waves varies in time with the rotation of Io. This seems to puzzle astronomers and a number of theories to explain it have been proposed which I am not sure I exactly understand. (I am not, after all, a professional astronomer.)

I suspect that these explanations must surely take into account Io's tidal action on Jupiter's huge atmosphere. This tidal action must affect the turbulence of that atmosphere and therefore its radio emission. In the extremely unlikely case that this has not been considered, I offer the suggestion to all comers free of charge.

That leaves only one more thing to consider. I have discussed the effect of the Sun upon Earth's tides. This is not terribly large (0·46) and one can expect that since the Sun's tide-producing effect drops off as the cube of the distance, the effect on planets more distant than the Earth would prove to be insignificant.

What about the effect on Venus and Mercury, however, which are closer to the Sun than is the Earth?

Well, according to my calculations, the Sun's tidal effect on Venus is 1·06 and its effect on Mercury is 3·77.

These are intermediate figures. They are more than the Moon's effect on Earth, which is not sufficient to stop Earth's rotation altogether; but they are less than the Earth's effect on the Moon, which was enough to stop the Moon.

One might suppose, then, that the rotations of Venus and Mercury, while slowed, would not yet have slowed to a stop.

Nevertheless, for a long time, the rotations of Venus and Mercury *were* considered as having been stopped, so that both planets faced a single side to the Sun at all times. In the case of Venus, this was a pure guess, for no one had ever seen the surface, but in the case of Mercury, where surface markings could be made out (though obscurely) the feeling seemed to check with observations.

In the last year or two, however, this view has had to be revised in the case of both planets. Venus and Mercury are each rotating slowly with respect to the Sun as (with the wisdom of hindsight) one might have suspected from the figures on tidal effects.

CHAPTER TWO

THE ROCKS OF DAMOCLES

In some ways, science fiction writers aren't doing too well these days. In late 1962, Mariner II seemed to settle the question of the surface temperature of Venus, placing it far above the boiling point of water.

With that, there vanished some of the most beautiful settings for s.f. stories. Old-timers may remember with nostalgia, as I do, the moist, swampy world of Weinbaum's 'Parasite Planet'. Well, it's gone! For that matter, I wrote a short novel under a pseudonym, a number of years ago, that was set on a Venus that was one huge ocean, with Earth-cities built underwater in the shallower regions[1] . . . All gone!

Now along comes Mariner IV and discovers craters (but no canals) on Mars.

No one expected that! I don't know of a single science fiction story that had ever placed craters on Mars . . . Canals, yes, but craters, no! I have written several stories set on Mars and I have always mentioned the canals (I placed no water in them; I knew enough for that) but I've never had craters.

And yet, going by the photographs sent back by various Mariner probes, the Martian surface is, in some places anyway, at least as rich in craters as the Moon is.

Fortunately for science-fictional self-respect, astronomers themselves didn't do much better. Not one of them, as far as I know, suggested Venus might be as hot as Mercury until after the first microwave observations came to be analysed in detail. And very few even speculated on the possibility of a cratered Mars.

In the first flush of the Martian pictures, the newspapers announced that this meant there was no life on Mars. To be sure, the pictures didn't give us life-enthusiasts anything to cheer about but, as it turns out, things aren't all that bad.

The mere fact that the pictures show no signs of life means nothing in itself, of course. Some of our own weather satellites have taken numerous pictures of the Earth under conditions comparable to those of Mariner IV and Mars, and the Earth pictures show no signs of life on our planet, either. I don't mean no signs of man; I mean no signs of any life at all. And, mind you, we know where to look for signs of life on Earth, and what to look for.

A more subtle argument for the anti-life view rests upon the mere existence of all those craters on Mars. If Mars had ever had an ample ocean and atmosphere those craters would have been eroded away. Since the craters are there, goes the argument, Mars has always been desiccated and almost airless and, therefore, the chances of life having developed in the first place are extremely small.

Quickly, however, the pro-life forces shot back. Since Mars is much closer to the asteroid zone than the Moon is, and since the asteroids are very likely to be the source of the large bodies that collide with planets and form sizable craters, Mars ought to have something like twenty-five times as many craters per unit surface area as the Moon does. It appears to have considerably fewer than that many. What happened to the other five-sixths?

Eroded away!

If so, then the craters we do see represent the youngest ones, the ones that have not had time to erode away yet.

Assuming that the process of erosion proceeds at uniform speed, then the marks we see only tell us about the last sixth or so of Martian history — say six to seven hundred million years.

What happened before that we still can't tell. Water may have been present in larger quantities before then and life might have started on a comparatively water-rich and water-comfortable Mars. If so, Martian life may have been hardy enough to survive even now on the gradually bleaching bones of the planet.

Maybe not, but we can't tell from only the pictures we have. We will need much more detailed photographs or, better yet, a manned expedition to Mars.[2]

Still, craters on Mars do set one thinking. In the course of the history of the solar system (its inner regions, at least) the major worlds must have undergone a continuous peppering bombardment from smaller bodies. The Moon and Mars bear the visible scars of this and it is quite out of the question to suppose that the Earth could possibly have escaped its share of the bombardment.

Although the Earth is as far from the asteroid zone as the Moon is, it is eighty-one times as massive as the Moon and, at equivalent distances, pulls with eighty-one times the force. Furthermore, Earth is the larger target, with fourteen times the cross-sectional area of the Moon, so it should have endured many times as many collisions. In fact, although Earth is much farther from the asteroid zone than Mars is, it has 3·5 times the cross-sectional area and ten times the mass of Mars, and I suspect it has suffered more of a peppering than Mars has.

The Moon is supposed to have 300,000 craters with diameters of one kilometre or more. Even allowing for the fact that the Earth is seventy per cent ocean, which can absorb collisions without being marked up as a result, the remaining 30 per cent — Earth's land surface — might well have suffered at least a million collisions in its billions of years of history.

Where are they all? Erased! The effect of wind, water, and living things quickly wipes them out and every last trace of more than 99 per cent of the craters formed on Earth must have vanished by now.

But surely there are remnants of the more recent ones. All one need do is look for depressions in the Earth that are more or less circular. These are easy to find, as a matter of fact, especially since they need only be roughly circular. After all, unevennesses can be put down to the effects of erosion, slippage, further bombardment, and so on.

A nice near-circular depression might fill with water and form a near-circular sea. The Aral Sea is an example. Then, too, the northern boundary of the Black Sea forms a near-circular arc. The Gulf of Mexico can almost be made to fit a semicircle. For that matter the Indian Ocean, and even the Pacific Ocean, have coastlines that can be made to fit circles with surprising closeness.

Whether you find such circles or not depends on how eagerly you want to find them and how ready you are to dismiss departure from strict circularity.

One of those most anxious to find possible craters is Dr. Frank Dachille of Pennsylvania State University. At least, the university has just sent me a discussion of his views, which includes a map on which no less than forty-two 'probable and putative' meteorite craters or groups of craters are listed in the United States and near-vicinity alone.

Of these, the largest in the United States proper is what is marked down as the 'Michigan Basin'. This is the near-circle formed by Lakes Michigan and Huron, a circle some three hundred miles in diameter. According to Dachille's estimate a crater this large would have to be caused by a meteorite thirty or forty miles in diameter.

On the map a still larger crater is indicated as 'Kelly Crater', and this marks out the Atlantic coastline of the United States along the continental shelf. This forms an arc of about one-third of a circle which would be over twelve hundred miles in diameter if it were complete. I suppose that would require a meteorite a hundred miles in diameter or so — in other words, one of the larger asteroids.[3]

One might dismiss such catastrophic events as having been confined to a highly specialised period of planetary history. At the very beginning of the formation of the solar system, planetesimals were gathering together to form the planets and the last few really large ones could have left the huge scars that mark major formations on the Earth and Moon.

Or else there was a period in planetary history during which a possible planet between Mars and Jupiter exploded (or underwent a series of explosions) leaving the asteroid belt behind and riddling the solar system from Mercury to Jupiter with flying shrapnel in bits up to a hundred miles across.

In either case, one might argue, this specialised period is over, the damage is done, the craters are formed, and we can dismiss the matter. There are no more planet-busters, no more asteroids of one hundred miles in diameter or more floating around within reaching distance of the Earth.

Indeed, that is so. There is no body that ever approaches within 25 million miles of the Earth that is over, let us say, twenty-five miles in diameter, except for the Moon itself, and there is no reason to expect the Moon to leave its orbit. Are we safe, then?

No, we are not! Space is loaded with dust particles and pebbles that burst into our atmosphere, glow, and vaporise harmlessly. In addition, however, larger chunks — not large enough to gouge out an ocean, perhaps, but large enough to do horrendous damage — are moving near us.

We can after all find craters that are indubitably craters and that are in such good condition that they must be quite new. The most spectacular of these craters is located near Winslow, Arizona. It looks just like a small lunar crater; it is roughly circular, with an average diameter of about 4,150 feet or four-fifths of a mile. It is 570 feet deep and the bottom is filled with a layer of mashed and broken debris about 600 feet thick. It is surrounded by a wall which is from 130 to 160 feet higher than the surrounding plain.

The first to demonstrate that the crater was caused by a meteorite (and was not an extinct volcano) was the American mining engineer Daniel Moreau Barringer, and the site is therefore called the Barringer Crater. Because of its origin it also bears the more dramatic name Meteor Crater. I have even seen it called the Great Barringer Meteor Crater, a name which honours, at one stroke, the size, the man, and the origin.

The fact that the strike took place in an arid region where the effects of water and living things are minimal has kept the Barringer Crater in better preservation than would have been the case if it were located in most places on Earth. Even so, its condition is such that it is not likely to be more than fifty thousand years old and, geologically speaking, that is yesterday.

If the meteorite that formed it (some millions of tons in mass) had struck now, and in the proper spot, it could, at one stroke, wipe out the largest city on Earth, and more or less destroy vast tracts of its suburbs.

Other strikes (not as large to be sure) have taken place even in historic times, two respectable ones in the twentieth century. One took place in central Siberia in 1908. It involved a meteor of only a few dozen tons of mass, perhaps, but that was enough to gouge out craters up to one hundred and fifty feet in diameter and to knock down trees for twenty to thirty miles around.[4]

That will do, and how! A fall like that in the middle of Manhattan would probably knock down every building on the island and large numbers across the rivers on either side, killing several million people within minutes of impact.

In fact, that 1908 fall did come close to wiping out a major city. It has been calculated that if the meteor had moved in an orbit parallel to its actual path, but had been displaced just enough in space to allow the Earth to rotate for five more hours before impact, it would have hit St. Petersburg (then the capital of the Russian Empire and now Leningrad) right on the nose.

Then, in 1947, there was another such fall, but a smaller one, in far-eastern Siberia.

Two falls, then, both in Siberia, and both doing virtually no damage except to trees and wild animals. Mankind has clearly had an unusual run of luck.

There are some astronomers who estimate that there may be two such 'city-busters' hitting Earth per century. If so, we can make some calculations. The area of New York City is about $1/670,000$ of the Earth's surface. If we assume that a city-buster can strike any place on Earth, at random, then a given city-buster has 1 chance in 670,000 of hitting New York.

If it comes, on the average, once every fifty years, then its chances of hitting New York in any one particular year is 1 in 670,000 times 50, or 1 in 33,000,000.

But New York City is only one city out of many. If we consider that the total densely urbanised area on Earth is 330 times the area of New York City (my spur-of-the-moment estimate) then the chance of some urban centre being flattened by a city-buster in any given year is 1 in 100,000.

To put it another way, it's even money that sometime within the next hundred thousand years, some good-sized city somewhere on Earth will be wiped out by a city-buster. And that is probably an over-optimistic prediction since the urbanised area on Earth is increasing and may continue to increase for quite a while, presenting a much better target.

This makes it clear, too, why no urbanised area in the past has been destroyed. Cities have only existed for say seven thousand years and until the last couple of centuries, the really large ones have been few and widely scattered. The chances for a major disaster of this sort having taken place at any time in recorded history is probably not better than 1 in 100 — and it hasn't happened.

But must we have a direct hit? What about the dangers of a near-miss. On land, a near-miss may be tolerable. Even a 1908-type meteorite striking land fifty miles from a populated centre may leave that centre intact. But what if the meteorite strikes the ocean? Three out of four, after all, ought to.

If the meteorite is not inordinately large and if it strikes far enough away from a coastline, the damage may be small. But there is always a chance that a large meteorite may strike near a coastline, even perhaps in a landlocked arm of the sea. It might then do severe damage, and since the appropriate regions of the sea are much larger in area than are the cities of the world, the chances of a catastrophic oceanic near-miss are correspondingly greater than the chances of a dead-centre hit on a city.

The oceanic near-miss might be expected, on the average, not once in a hundred thousand years, but once in ten thousand years or even less. In short, there should be some record of such disasters in historical times, and I think that perhaps there are.

Noah's Flood *did* happen. Or, put it this way — There

was a vast and disastrous flood in the Tigris-Euphrates some six thousand years ago. Concerning this, Babylonian stories were handed down which, with the generations, were interlarded with mythological detail. The tale of Noah's Flood as given in the Bible is a version of those stories.

Nor is this speculation. Archaeological probings on the sites of some of the ancient cities of Babylonia have come across thick layers of sediment within which there are no human remains or artifacts.

What laid down these sediments? The usual suggestion is that since the Tigris-Euphrates complex floods occasionally, it may have flooded particularly disastrously on one occasion. This has always seemed insufficient to me. I don't see how any river flood can possibly lay down all the observed sediment or do the kind of damage dramatised later in the tales of universal flood, death, and destruction — even allowing for the natural exaggeration of storytellers.

I have an alternative suggestion which, as far as I know, is original with me. I have seen nothing about it anywhere.

What if a city-buster meteorite had, some six thousand years ago, landed in the Persian Gulf. The Persian Gulf is nearly landlocked and such an impact might well have heaved up a wall of water that would move northwestward and would burst in, with absolutely devastating impact, upon the low plain of the Tigris-Euphrates valley.

It would be a super-tsunami, a tidal wave to end all tidal waves, and it would scour much of the valley clean. The water would cover what was indeed 'all the world' to the inhabitants and drown countless numbers in its path.

In support of this notion, I would like to point out that the Bible speaks of more than rain. Genesis 7: 11 says not only that 'the windows of heaven were opened' meaning that it rained, but also that 'the same day were all the fountains of the great deep broken up'. Meaning what? Meaning, it seems to me, that the water came in from the sea.

Furthermore, Noah's ark lacked any motive power, either sails or oars, and simply drifted. Where did it drift? It came to rest on the mountains of Ararat (and ancient Urartu) in the Caucasian foothills north-west of the Tigris-Euphrates. But an ordinary river flood would have washed boats south-eastward out to sea. Only a tidal wave of unprecedented scope would have carried the ark northwestward.[5]

Nor need Noah's Flood be the only tale of a remembered oceanic near-miss. Many peoples (but not all) have flood legends, and it may have been such a flood legend that gave rise to the dim tales that Plato finally dramatised in his story of Atlantis.[6] Such catastrophes may indeed have happened more than once in the memory of man.

When will the next disaster come? A hundred thousand years hence? A thousand years? Tomorrow? There's no way of telling.

Of course, we might scan near-space and see what's floating around there.

Till 1898, the answer was Nothing! Between the orbits of Mars and Venus was nothing of any consequence save the Earth and Moon, for all anyone could tell. Nothing worse than pebbles and small boulders, at any rate, that could produce the effect of a shooting star and occasionally reach Earth's surface. A meteorite might conceivably kill a man or demolish a house, but the damage to be expected of meteorites was only the tiniest fraction of that done by lightning bolts, for instance, and man has managed to live with the thunderstorm.

Then in 1898, the German astronomer Gustav Witt discovered Asteroid 433. Nothing unusual, until Witt calculated its orbit. This was elliptical, of course, and in that part of the elliptical path that was farthest from the Sun, the asteroid travelled between Mars and Jupiter as did all asteroids till then discovered.

In the remainder of its orbit, however, it travelled between the orbits of Mars and Earth. Its orbit approached within 14 million miles of that of Earth and at set intervals, when both objects were at just the right place in their respective orbits, that approach would be realised and the asteroid would be at only half the distance from us that Venus is. Witt named the asteroid Eros after the son of Mars and Venus in the classical myths. That began the practice of masculine names for all asteroids with unusual orbits.

Eros's close approach was a matter of self-congratulations among scientists. It could be used to determine the dimensions of the solar system with unprecedented accuracy, when close. And this was done in 1931, when it approached within 17 million miles. From the periodic flickering of its light, it was decided that Eros was not a sphere but an irregular, roughly brick-shaped object which brightened when we saw it long-ways and dimmed when we saw it end-ways. Its longest diameter is estimated to be fifteen miles and its shortest five miles.

There was no particular nervousness felt about Eros's close approach. After all, 14 million miles isn't exactly *close*, you know.

As time went on, however, several additional objects were discovered with orbits that came closer to Earth's than that of Venus does, and such objects came to be called 'Earth-grazers'. Several of them passed closer to the Sun than Venus does and one of them, Icarus, actually moves in closer to the Sun than Mercury does.

The climax came in 1937, when the asteroid Hermes was discovered by an astronomer named Reinmuth. On October 30th, it passed within 487,000 miles of the Earth and the calculation of its orbit seems to show that it might possibly come within 200,000 miles — closer than the Moon! (For information on some Earth-grazers, *see* Table 2.)

Table 2

THE EARTH-GRAZERS[1]

NAME	YEAR OF DIS- COVERY	LENGTH OF ORBITAL PERIOD (YEARS)	ESTIMATED MAXIMUM DIAMETER (MILES)	ESTIMATED MASS (TRILLIONS OF TONS)	CLOSEST APPROACH TO EARTH (MILLIONS OF MILES)
Albert	1911	c. 4	3	300	20
Eros	1898	1·76	15	15,000	14
Amor	1932	2·67	10	12,000	10
Apollo	1932	1·81	2	100	7
Icarus	1949	1·12	1	12	4
Adonis	1936	2·76	1	12	1·5
Hermes	1937	1·47	1	12	0·2

Even so, why worry? A miss of two hundred thousand miles is still a fair-sized miss, isn't it?

No, it isn't, and for three reasons. In the first place, the orbits of asteroids aren't necessarily fixed. The Earth-grazers have small masses, astronomically speaking, and a close approach to a larger world can introduce changes in that orbit. Comets, for instance, which have asteroidal masses, have been observed on more than one occasion to undergo radical orbital changes as a result of close approaches to Jupiter. Hermes doesn't approach Jupiter at all closely to be sure, but it does skim the Earth-Moon system, and, on occasion, Mercury, and it is subject to orbital change for that reason.

As a matter of fact, Hermes hasn't been located since its first sighting in 1937, though it should have come fairly close every three years or so. This may well mean that its orbit has already changed somewhat so that we don't know the right place to look for it and will rediscover it, if at all, only by accident.

A random change in Hermes' orbit is much more likely to move it away from Earth than towards it, since there is much more room away than towards. Still, there is a finite chance that such a change might cause it to zero in, and a direct collision of Hermes and Earth is horrifying to think of. The Earth won't be perceptibly damaged, but we could be. If you think that a meteorite a few million tons in mass could gouge out a hole nearly a mile across, you can see that Hermes, with a few trillions of tons of mass, could excavate a good-sized portion of an American state or of a European nation.

Secondly, these Earth-grazers probably haven't been in their orbit through all the history of the solar system. Some collision, some perturbation, has shifted their orbits — originally within the asteroid belt — and caused them to fall in towards the Sun. On occasion, then, new ones might join them. Naturally, it is the smaller asteroids that stand the best chance of serious orbital changes but there are thousands of Hermes-sized asteroids in the asteroid belt and Earth can still witness the coming of unwelcome new strangers.

Thirdly, the only Earth-grazers we can detect are those big enough to see at hundreds of thousands of miles. There are bound to be smaller ones in greater profusion. If there are half a dozen objects a mile in diameter and more that come wandering into near-space on occasion, there may be half a thousand or more which are one hundred feet in diameter or so, and which could still do tremendous damage if they wandered in too closely.

No, I see no way, at present, of either predicting or avoiding the occurrence of a very occasional major catastrophe. We spin through space with the possibility of collision with these rocks of Damocles ever present.

In the future, perhaps, things may be different. The men in the space stations that will eventually be set up about the Earth may find themselves, among other things, on the watch for the Earth-grazers, something like the iceberg watch conducted in northern waters since the sinking of the Titanic (but much more difficult of course).

The rocks, boulders, and mountains of space may be painstakingly tagged and numbered. Their changing orbits may be kept under steady watch. Then, a hundred years from now, perhaps, or a thousand, some computer on such a station will sound the alarm: 'Collision orbit!'

Then a counter-attack, kept in waiting for all that time would be set in motion. The dangerous rock would be met with an H-bomb (or, by that time, something more appropriate) designed to trigger off on collision. The rock would glow and vaporise and change from a boulder to a conglomeration of pebbles.

Even if they continued on course, the threat would be lifted. Earth would merely be treated to a spectacular (and harmless) shower of shooting stars.

Until then, however, the Rocks of Damocles remain suspended, and eternity for millions of us may, at any time, be an hour away.

CHAPTER THREE

HARMONY IN HEAVEN

I never actually took any courses in astronomy, which is something I regret for, looking back on it now, there were a number of courses I did take which I might cheerfully have sacrificed for a bit of astronomy.

However, one must look at the bright side, which is that now, every once in a while, I come across a little item in my astronomical reading which gladdens my heart by teaching me something new. If I had had formal training in the field, then these items would all have been old stuff and I would have missed my moments of delight.

For instance, I have come across a recent text in astronomy, *Introduction to Astronomy* by Dean B. McLaughlin[1] (Houghton Mifflin, 1961) which has delighted me in this fashion in several places. Let me, therefore, recommend it to all of you without reservation.

As an example, Professor McLaughlin intrigued me so much with his comments on Kepler's harmonic law that, in my ecstasy, I devoted more thought to it than I had ever done before, and I see no reason why I should not share the results of that thinking with you. In fact, I insist upon it.

I might begin, I suppose, by answering the question that I know is in all your minds: What is Kepler's harmonic law? Well —

In 1619, the German astronomer, Johannes Kepler, discovered a neat relationship between the relative distances of the planets from the Sun, and their periods of revolution about the Sun.

Now for two thousand years, philosophers had felt that planets were spaced at such distances that their movements gave rise to sounds that united in heavenly harmony (the 'music of the spheres'). This was in analogy to the manner in which strings of certain different lengths gave forth sound that united in pleasing harmony when simultaneously struck.

For that reason, Kepler's relationship of distances and periods, which is usually called, with scientific dullness, 'Kepler's third law' (since he had earlier discovered two other important generalisations about planetary orbits) is also called, much more romantically, 'Kepler's harmonic law'.

The law may be stated thus: 'The squares of the periods of the planets are proportional to the cubes of their mean distances from the Sun.'

To follow up the consequences of this, let's get slightly mathematical (as slightly as possible, I promise). Let's begin by considering two planets of the solar system, planet-1 and planet-2. Planet-1 is at mean distance D_1 from the Sun and planet-2 is at mean distance D_2. ('Mean' means 'average'.) Their periods of revolution are, respectively, P_1 and P_2. Then, by Kepler's harmonic law, we can say that:

$$P_1^2/P_2^2 = D_1^3/D_2^3 \qquad \text{(Equation 1)}$$

This is not a very complicated equation, but any equation that can be simplified *should* be simplified, and that's what I'm going to do next. Let's pretend that planet-2 is the Earth and that we are going to measure all periods of revolution in years and all distances in astronomical units (A.U.).

The period of revolution of the Earth, by definition, is one year; therefore P_2 and $P_2{}^2$ both equal 1. Then, too, the astronomical unit is defined as the mean distance of the Earth from the Sun. Consequently, the Earth is 1 A.U. from the Sun, which means that D_2 and $D_2{}^3$ both equal 1.

The denominators of both fractions in Equation 1 become unity and disappear. With only one set of P's and D's to worry about, we can eliminate subscripts and write Equation 1 simply as follows:

$$P^2 = D^3 \qquad \text{(Equation 2)}$$

provided we remember to express P in years and D in astronomical units.

Just to show how this works let's consider the nine major planets of the solar system, and list for each the period of revolution in years and the distance from the Sun in astronomical units (*see* Table 3). If, for each planet, you take the

Table 3

PLANET	P (PERIOD OF REVOLUTION IN YEARS)	D (MEAN DISTANCE IN A.U.)
Mercury	0·241	0·387
Venus	0·615	0·723
Earth	1·000	1·000
Mars	1·881	1·524
Jupiter	11·86	5·203
Saturn	29·46	9·54
Uranus	84·01	19·18
Neptune	164·8	30·06
Pluto	248·4	39·52

square of the value under P and the cube of the value under D you will find, indeed, that the two results are virtually identical.

Of course, the period and distance of any given planet can be determined separately and independently by actual observation. The connection between the two, therefore, is interesting but not vital. However, what if we can't determine both quantities independently. Suppose, for instance, you imagined a planet between Mars and Jupiter at a distance of just 4 A.U. from the Sun. What would its period of revolution be? Or if you imagined a far, far distant planet, 6,000 A.U. from the Sun. What would *its* period of revolution be?

From Equation 2, we see that:

$$P = \sqrt{D^3} \qquad \text{(Equation 3)}$$

and therefore we can answer the questions easily. In the case of the planet between Mars and Jupiter, the period of revolution would be the square root of the cube of four, or just eight years. As for the far distant planet, its period would be the square root of the cube of 6,000 and that comes to 465,000 years.

You can work it the other way round, too, by converting Equation 2 into:

$$D = \sqrt[3]{P^2} \qquad \text{(Equation 4)}$$

You can then find out how distant from the Sun a planet must be to have a period of revolution of just twenty years, of just one million years.

In the former case, you must take the cube root of the square of twenty and in the latter, the cube root of the square of one million. This gives you an answer of 7·35 A.U. for the first case and about 10,000 A.U. for the second.

We can have a little fun, now, by seeking extremes. For instance, how far out can a planet be and still be a member of the solar system? The nearest star system to ourselves is Alpha Centauri, which is 4·3 light-years away. Any planet which is as close as 2 light-years to the Sun must therefore be closer to the Sun than to any other star no matter what the plane of its orbit. It is safely in the Sun's grip and let's consider it the 'farthest reasonable planet'.

An astronomical unit is equal to about 93,000,000 miles while a light-year is equal to about 5,860,000,000,000 miles. Therefore, one light-year is equal to about 63,000 A.U. and our farthest reasonable planet is at a distance of about 126,000 A.U. From Equation 3, then, we can see that the period of the farthest reasonable planet is about 45,000,000 years.

Let's ask next, how close a planet can be to the Sun? Let's ignore temperature and gas resistance and suppose that a planet can circle the Sun at its equator, just skimming its surface. We can call this a 'surface planet'.

The distance of a planet from the Sun is always measured centre to centre. If we consider the surface planet to be of negligible size, then its distance from the Sun is equal to the radius of the Sun, which is 432,300 miles or 0·00465 A.U. Again using Equation 3, we can show that the period of such a body is 0·00031 years or 2·73 hours.

Next let's find out how fast a planet is moving, on the average, in miles per second (relative to the Sun). To do so, let's first figure out how many seconds it takes the planet to make a complete turn in its orbit. We already have that period in years (P). In each year, there are about 31,557,000 seconds. Therefore the period of the planet in seconds is 31,557,000 P.

An astronomical unit is, as I said before, about 93,000,000 miles. We have the distance of a planet in astronomical units (D), so that the distance in miles is 93,000,000 D. What we really need at this point, however, is the length of the orbit itself. If we assume the orbit to be an exact circle (which is approximately true) then its length is equal to its distance from the Sun, multiplied by twice 'pi'. The value of 'pi' is 3·1416 and twice that is 6·2832. If we multiply that by the distance of the planet in miles, we get the length of the planetary orbit in miles, and that is 584,000,000 D.

To find the average velocity of a planet in miles per second, we must divide the length of the orbit in miles (584,000,000 D) by the duration of its period of revolution in seconds (31,557,000 P). This gives us the value 18·5 D/P, for the mean orbital velocity of a planet.

We can simplify this by remembering that $P = \sqrt{D^3}$, according to Equation 3, so that we can write the velocity of a moving planet as $18·5 \, D/\sqrt{D^3}$. Since $\sqrt{D^3}$ is equal to $\sqrt{D^2 \times D}$, or $D\sqrt{D}$, we can write the velocity of a planet in orbit as equal to $18·5 \, D/D\sqrt{D}$ or, in a final simplification, letting V stand for velocity:

$$V = 18·5/\sqrt{D} \qquad \text{(Equation 5)}$$

Remember that D represents the distance of a planet from the Sun in astronomical units. For the Earth the value of D is equal to 1 and the square root of D is also equal to 1. Therefore, the Earth moves in its orbit at the average rate of 18·5 miles per second.

Since D is known for the other planets, the mean orbital velocity can be calculated without trouble by taking the square root of D and dividing it into 18·5. The result is Table 4.

Nowadays, velocity is often spoken of in 'Mach numbers', where Mach 1 is equivalent to the speed of sound in air, Mach 2 to twice that speed, and so on. At 0° C. the speed of sound is 1,090 feet per second, or just about 0·2 miles per second. Our fastest aeroplanes are now moving along at Mach 2 and more, while an astronaut in orbit moves at about Mach 25 with respect to the Earth.

Compare this with Pluto, which moves (with respect to the Sun) at a mere Mach 14·5, only half the velocity of an astronaut. The Earth on the other hand is moving at a respectable Mach 93 and Mercury at a zippy Mach 149.

But let's try our extremes again.

Table 4

PLANET	MEAN ORBITAL VELOCITY (MILES PER SECOND)
Mercury	29·8
Venus	21·7
Earth	18·5
Mars	15·0
Jupiter	8·2
Saturn	6·0
Uranus	4·2
Neptune	3·4
Pluto	2·9

The farthest reasonable planet, at 126,000 A.U., would have an orbital velocity of just about 0·052 miles per second, or about Mach 0·26. It's rather impressive that even at a distance of two light-years, the Sun is still capable of lashing a planet into travelling at one-quarter the speed of sound.

As for the surface planet at a distance of 0·00465 A.U., its orbital velocity must be 271 miles per second, or Mach 1,355. (Incidentally, the fastest conceivable velocity, that of light in a vacuum, is equal to about Mach 930,000, so watch out for anyone who talks casually about Mach 1,000,000. Bet him you can't reach Mach 1,000,000 and you'll win.)

Actually, a planet orbits about the Sun not in a circle but in an ellipse with the Sun at one focus (Kepler's first law). If you imagine a line connecting the Sun and the planet (a 'radius vector'), that line would sweep out equal areas in equal times. (This is Kepler's second law.) When the planet is close to the Sun, the radius vector is short and, to sweep out a given area, it must move through a comparatively large angle. When the planet is far from the Sun, the radius vector is long and, to sweep out the same area, needs to move through a smaller angle.

Thus, Kepler's second law describes the manner in which a planet's orbital velocity speeds up as it approaches the Sun and slows down as it recedes from it. I would like to point out one consequence of this without going into mathematical detail.

Imagine a planet suddenly increasing its velocity at some point in its orbit. The effect upon it would be analogous to that of throwing it away from the Sun. It would move away from the Sun at a steadily decreasing velocity, come to a halt, and then start falling towards the Sun again.

This resembles the situation where one throws a stone into the air here on Earth, but since the planet is also revolving about the Sun, the effect is not a simple up-and-down motion, as it is in the case of the stone.

Instead, the planet revolves as it recedes from the Sun, its orbital velocity decreasing until it reaches a point that is precisely on the other side of the Sun from the point at which its velocity had suddenly increased. At this point on the other side of the Sun, its distance from the Sun has increased to a maximum (aphelion), and its orbital velocity has slowed to a minimum.

As the planet continues past the aphelion, it begins to approach the Sun again and its orbital velocity increases once more. When it returns to the place at which it had suddenly increased its velocity, it would be at that point in its new orbit which was nearest the Sun (perihelion) and its orbital velocity would then be at a maximum.

The greater the velocity at a given perihelion distance, the more distant the aphelion and the more elongated the elliptical orbit. The elongation increases at a greater and greater rate with equal increments of speed because as the aphelion recedes, the strength of the Sun's gravity weakens and it can do less and less to prevent a still further recession.

Eventually, at some particular velocity at a given perihelion distance, the ellipse elongates to infinity — that is, it becomes a parabola. The planet continues along the parabolic orbit, receding from the Sun forever and never returning. This velocity is the 'escape velocity' and it can be determined for any given planet by multiplying the mean orbital velocity in Table 4 by the square root of two; that is, by 1·414. The result is shown in Table 5.

Table 5

PLANET	ESCAPE VELOCITY (MILES PER SECOND)
Mercury	42·1
Venus	30·7
Earth	26·2
Mars	21·2
Jupiter	11·6
Saturn	8·5
Uranus	5·9
Neptune	4·8
Pluto	4·1

Thus, if the Earth, for any reason, ever moved at 26·2 miles per second or more, it would leave the solar system forever. (However, don't lose sleep over this. There is nothing, short of the invasion of another star, that can bring this about.)

Escape velocity for the farthest reasonable planet would

be 0·073 miles per second while that for the surface planet would be 385 miles per second.

Isaac Newton used Kepler's three laws as a guide in the working out of his own theory of gravitation. Once gravitation was worked out, Newton showed that Kepler's three laws could be deduced from it. In fact, he showed that Kepler's harmonic law as originally stated (*see* Equation 1) was only an approximation. In order to make it really exact, the masses of the Sun and the planets had to be taken into account. Equation 1 would have to be written in this way:

$$(M+m_1)P_1^2/(M+m_2)P_2^2 = D_1^3/D_2^3 \qquad \text{(Equation 6)}$$

where, as before, P_1 and P_2 are the periods of revolution of planet-1 and planet-2, D_1 and D_2 are their respective distances, and where the new symbols m_1 and m_2 are their respective masses. The symbol M represents the mass of the Sun.

As it happens, the mass of the Sun is overwhelmingly greater than the mass of any of the planets. Even the largest planet, Jupiter, has only $1/1000$ the mass of the Sun. Consequently, the sum of M and m_1 or of M and m_2 can be taken, without significant inaccuracy, to be equal to M itself. Equation 6 can therefore be written as follows:

$$MP_1^2/MP_2^2 = D_1^3/D_2^3 \qquad \text{(Equation 7)}$$

The M's cancel and we have Equation 1.

Of course, you may decide that since Newton's correct form works out to be just about exactly that of Kepler's approximate form, why not stick with Kepler, who is simpler?

Ah, but Newton's form can be applied more broadly.

Jupiter's satellites had been discovered nine years before Kepler had announced his harmonic law. Kepler had worked out that law entirely from the planets, yet when he studied

Jupiter's satellite system, he found it applied to that, too.

Newton was able to show from his theory of gravitation that all three of Kepler's laws would apply to any system of bodies moving about some central body and his form of the harmonic law could be applied to two or more different systems at once.

Suppose, for instance, that planet-1 is circling Sun-1 and planet-2 is circling Sun-2. You can say that:

$$(M_1 + m_1)P_1^2/(M_2 + m_2)P_2^2 = D_1^3/D_2^3 \qquad \text{(Equation 8)}$$

where M_1 and M_2 are the masses of Sun-1 and Sun-2, where m_1, P_1, and D_1 are the mass, period, and distance of planet-1, and where m_2, P_2, and D_2 are the mass, period, and distance of planet-2.

Now let's simplify that rather formidable assemblage of symbols. In the first place, we can take it for granted that the planet is always so much smaller than the Sun that its mass can be neglected. (This is not always true but it's true in the solar system.) In other words, we can eliminate m_1 and m_2 and write Equation 8 as:

$$M_1 P_1^2/M_2 P_2^2 = D_1^3/D_2^3 \qquad \text{(Equation 9)}$$

Secondly, let's take the situation of the Earth revolving about the Sun as the norm and consider it to be the planet-2/Sun-2 system. We will measure all distances in astronomical units so that D_2^3 will equal 1. We will measure all periods of revolution in years so that P_2^2 will equal 1. Also we will measure the mass of all Suns in terms of the mass of our own Sun taken as 1. That means that M_2, the mass of the Sun, is the equal to 1. Equation 9 becomes (dropping all subscripts)

$$MP^2 = D^3 \qquad \text{(Equation 10)}$$

where the symbols refer to the system other than the Earth/Sun system.

Suppose, for instance, that for the other Sun, we chose the Earth itself. (The Earth can serve as a central body around which smaller bodies, satellites, can revolve.) Suppose, further, that we wanted to calculate the period of revolution of a body circling the Earth at a mean distance of 237,000 miles. Since it is a period of revolution we are seeking, let us rewrite Equation 10 as:

$$P = \sqrt{D^3/M} \qquad \text{(Equation 11)}$$

The value of D is equal to 237,000 miles or 0·00255 A.U. The value of M is equal to the mass of the Earth expressed in Sun-masses. The Earth's mass is $1/_{332,500}$ of the Sun or 0·000003 Sun-masses. Substituting these values into Equation 11, we find that P, the period of revolution, comes out to 0·0745 years, or 27·3 days.

It happens that the Moon is at an average distance of 237,000 miles from the Earth, and it happens that its period of revolution (relative to the stars) is 27·3 days. Consequently, Kepler's harmonic law, as corrected by Newton, applies as much to the Earth-Moon system as to the Sun-planet system.

Furthermore, since the distance of the Moon from the Earth and the Moon's period of revolution are both known; and since the distance of the Earth from the Sun and the Earth's period of revolution are also both known; then if the mass of the Earth is known, the mass of the Sun can be calculated from Equation 9. Or, if the mass of the Sun is known, that of the Earth can be calculated.

The mass of the Earth was worked out by a method independent of the harmonic law in 1798. After that, the mass of any astronomical body which is itself at a known distance, and has a body circling it at a known distance and in a known period (all these quantities being easy to determine within the solar system) can quickly be determined. For this reason, the masses of Mars, Jupiter, Saturn, Uranus, and Neptune, all with satellites, are known with considerable accuracy.

The masses of Mercury, Venus, and Pluto, which lack known satellites, can only be worked out by more indirect means and are known with considerably less accuracy. (It seems unreasonable that the mass of Venus is less well known than that of Neptune when the latter is a hundred times farther from us, but now you see why.)[2]

The masses of the various satellites (except for the Moon itself, which is a special case) are hard to determine for similar reasons. The harmonic law can't be used because their masses are drowned by the much larger mass of their primary, and no other method of mass determination is as convenient or as accurate.

Periods, distances, and orbital velocities of satellites, real or imagined, can be worked out for any planet (real or imaginary) for which the mass is known, exactly as these quantities can be worked out for the planets with respect to the Sun.

Without going into arithmetical details, I will list some data in Table 5, on the surface satellite for each planet; the theoretical situation where a satellite just skims the planetary equator. For this, use must be made of both mass and radius of the planet and these values are so uncertain in the case of Pluto that I will leave it out. In its place, for comparison purposes, I include the Sun.

If you consider Table 6, you will see that the period of a minimum satellite can be long for either of two reasons. As in the case of Mercury, the planet is light and its gravitational force is so weak that the satellite is moved along slowly and takes several hours to negotiate even the small length of the planetary equator.

On the other hand, as in the case of the Sun or of Jupiter, the gravitational force is great and the surface satellite whizzes along at high speed, but the central body is so large that even at high speed, several hours must elapse before the circuit is completed.

Table 6

PLANET	SURFACE SATELLITE		
	PERIOD (HOURS)	PERIOD (MINUTES)	ORBITAL VELOCITY (MILES PER SEC.)
Mercury	3·13	188	1·87
Venus	1·44	86½	4·58
Earth	1·41	84½	4·95
Mars	1·65	99	2·27
Jupiter	2·96	177	26·4
Saturn	4·23	254	15·6
Uranus	2·62	157	9·85
Neptune	2·28	137	11·2
Sun	2·73	165	271

The period of the surface satellite is shortest when the planet packs as much mass as possible into as small a volume as possible. In other words, the greater the density of the central body, the shorter the period of the surface satellite. Since Saturn is the least dense of the bodies listed in Table 6, it is not surprising that its surface satellite has the longest period.

As it happens, of all the sizable bodies of the solar system, our own planet, Earth, is the densest. The period of its surface satellite is therefore the shortest.

An astronaut in orbit about the Earth, a hundred miles or so above the surface, is virtually a surface satellite and he completes his circuit of the earth in just under ninety minutes. An astronaut of no other sizable body in the solar system could perform so speedy a circumnavigation.

How's that for a system-wide distinction for Gagarin, Glenn, and company?

CHAPTER FOUR

THE TROJAN HEARSE

The very first story I ever had published (never mind how long ago that was)[1] concerned a spaceship that had come to grief in the asteroid zone. In it, I had a character comment on the foolhardiness of the captain in not moving out of the plane of the ecliptic (i.e. the plane of the earth's orbit, which is close to that in which virtually all the components of the solar system move) in order to go over or under the zone and avoid almost certain collision.

The picture I had in mind at that time was of an asteroidal zone as thickly strewn with asteroids as a beach was with pebbles. This is the same picture that exists, I believe, in the mind of almost all science-fiction writers and readers. Individual miners, one imagines, can easily hop from one piece of rubble to the next in search of valuable minerals. Vacationers can pitch their tents on one world and wave at the vacationers on neighbouring worlds. And so on.

How true is this picture? The number of asteroids so far discovered is just about 1,800, but, of course, the actual number is far higher. I have seen estimates that place the total number at 100,000.

Most of the asteroids are to be found between the orbits of Mars and Jupiter and within 30° of the plane of the ecliptic. Now, the total volume of space between those orbits and within that tilt to the ecliptic is (let's see now —

mumble, mumble, mumble) something like 200,000,000,000,-000,000,000,000,000 cubic miles. If we allow for a total quantity of 200,000 asteroids, to be on the safe side, then there is one asteroid for every 1,000,000,000,000,000,000,000 cubic miles.

This means that the average distance between asteroids is 10,000,000 miles. Perhaps we can cut that down to 1,000,000 miles for the more densely populated regions. Considering that the size of most asteroids is under a mile in diameter, you can see that from any one asteroid you will in all probability see no others with the naked eye. The vacationer will be lonely and the miner will have a heck of a problem reaching the next bit of rubble.

In fact, astronauts of the future will in all probability routinely pass through the asteroid zone on their way to the outer planets and never see a thing. Far from being a dreaded sign of danger, the occasional cry of 'asteroid in view' should bring all the tourists rushing to the portholes.[2]

Actually, we mustn't think of the asteroidal zone as evenly strewn with asteroids. There are such things as clusters and there are also bands within the zone that are virtually empty of matter.

The responsibility for both situations rests with the planet Jupiter and its strong pull on the other components of the solar system.

As an asteroid in the course of its motion makes its closest approach to Jupiter (in the course of *its* motion), the pull of Jupiter on that asteroid reaches a maximum. Under this maximum pull, the extent by which an asteroid is pulled out of its normal orbit (is 'perturbed') is also at a maximum.

Under ordinary circumstances, however, this approach of the asteroid to Jupiter occurs at different points in their orbits. Because of the rather elliptical and tilted orbits of most asteroids, the closest approach therefore takes place at varying angles, so that sometimes the asteroid is pulled forward at the time of its closest approach and sometimes backward, sometimes downward and sometimes upward. The net result is that the effect of the perturbations cancels out and that, in the long run, the asteroids will move in orbits that oscillate about some permanent average-orbit.

But suppose an asteroid circled about the sun at a mean distance of about 300,000,000 miles? It would then have a period of revolution of about six years as compared with Jupiter's period of twelve years.

If the asteroid were close to Jupiter at a given moment of time; then twelve years later, Jupiter would have made just one circuit and the asteroid just two circuits. Both would occupy the same relative positions again. This would repeat every twelve years. Every other revolution, the asteroid would find itself yanked in the same direction. The perturbations, instead of cancelling out, would build up. If the asteroid was constantly pulled forward at its close approach, it would gradually be moved into an orbit slightly more distant from the sun, and its year would become longer. Its period of revolution would then no longer match Jupiter's, and the perturbations would cease building up.

If, on the other hand, the asteroid were pulled backward each time, it would gradually be forced into an orbit that was closer to the sun. Its year would become shorter; it would no longer match Jupiter's; and again the perturbations would cease building up.

The general effect is that no asteroid is left in that portion of the zone where the period of revolution is just half that of Jupiter. Any asteroid originally in that portion of the zone moves either outward or inward; it does not stay put.

The same is true of that region of the zone in which an asteroid would have a period of revolution equal to four years, for then it would repeat its position with respect to Jupiter every three revolutions. If it had a period of revolution equal to 4·8 years, it would repeat its position with respect to Jupiter every five revolutions, and so on.

The regions of the asteroid zone which have thus been swept clear of asteroids by Jupiter are known as 'Kirkwood's gaps'. They are so named because the American astronomer Daniel Kirkwood called attention to these gaps in 1876 and explained them properly.

An exactly analogous situation is also to be found in the case of Saturn's rings — which is, in fact, why we speak of 'rings' rather than 'ring'.

The rings were first discovered by the Dutch astronomer Christian Huygens in 1655. To him it seemed a simple ring of light circling Saturn but touching it nowhere. In 1675, however, the Italian-born French astronomer Giovanni Domenico Cassini noticed that a dark gap divided the ring into a thick and bright inner portion and a thinner and somewhat less bright outer portion. This gap, which is 3,000 miles wide, has been called the 'Cassini division' ever since.

In 1850, a third, quite dim ring, closer to Saturn than are the others, was spied by the American astronomer George Phillips Bond. It is called the 'crape ring' because it is so dim. The crape ring is separated from the inner bright ring by a gap of 1,000 miles.

In 1859, the Scottish mathematician Clerk Maxwell showed that from gravitational considerations, the rings could not be one piece of material but had to consist of numerous light-reflecting fragments that seemed one piece only because of their distance. The fragments of the crape ring are much more sparsely distributed than those of the bright rings, which is why the crape ring seems so dim. This theoretical prediction was verified when the period of revolution of the rings was measured spectroscopically and found to vary from point to point. After all, if the rings were one piece, the period of revolution would be everywhere the same.

The innermost portion of the crape ring is a mere 7,000 miles above Saturn's surface. Those particles move most rapidly and have the shortest distance to cover. They revolve about Saturn in about $3\frac{1}{4}$ hours.

As one moves outward in the rings, the particles move more slowly and must cover greater distances, which means that the period of revolution mounts. Particles at the outermost edge of the rings have a period of revolution of about $13\frac{1}{2}$ hours.

If particles were to be found in Cassini's division, they would circle Saturn in a period of a little over 11 hours. But particles are not found in that region of the rings, which is why it stands out black against the brightness on either side. Why?

Well, outside the ring system, Saturn possesses a family of nine satellites, each of which has a gravitational field that produces perturbation in the motion of the particles of the rings. Saturn's innermost satellite, Mimas,[3] which lies only 35,000 miles beyond the outer edge of the rings, has a period of revolution of $22\frac{1}{2}$ hours. Enceladus, the second satellite, has a period of 33 hours; and Tethys, the third satellite, a period of 44 hours.

Any particles in Cassini's division would have a period of revolution half that of Mimas, a third that of Enceladus, and a fourth that of Tethys. No wonder the region is swept clean. (Actually, the satellites are small bodies and their perturbing effect is insignificant on anything larger than the gravel that makes up the rings. If this were not so, the satellites themselves would by now have been forced out of their own too-closely matching orbits.)

As for the gap between the crape ring and the inner bright one, particles within it would circle Saturn in a little over seven hours, one-third the period of revolution of Mimas and one-sixth that of Tethys. There are other smaller divisions in the ring system which can be explained in similar fashion.

I must stop here and point out a curiosity that I have never seen mentioned. Books on astronomy always point out that Phobos, the inner satellite of Mars, revolves about Mars in less time than it takes Mars to rotate about its axis. Mars' period of rotation is $24\frac{1}{2}$ hours while Phobos' period of revolution is only $7\frac{1}{2}$ hours. The books then go on to say that Phobos is the only satellite in the system of which this is true.

Well, that is correct if we consider natural satellites of appreciable size. However, each particle in Saturn's rings is really a satellite, and if they are counted in, the situation changes. The period of rotation of Saturn about its axis is $10\frac{1}{2}$ hours, and every particle in the crape ring and in the inner bright ring revolves about Saturn in less time than that. Therefore, far from there being only one satellite of the Phobos type, there are uncounted millions of them.

In addition, almost every artificial satellite sent up by the United States and the Soviet Union revolves about the earth in less than twenty-four hours. They, too, are of the Phobos type.

Gravitational perturbations act not only to sweep regions clear of particles but also to collect them. The most remarkable case is one where particles are collected not in a zone, but actually in a point.

To explain that, I will have to begin at the beginning. Newton's law of universal gravitation was a complete solution of the 'two-body problem' (at least in classical physics, where the modern innovations of relativity and quantum are ignored). That is, if the universe contains only two bodies, and the position and motion of each are known, then the law of gravitation is sufficient to predict the exact relative positions of the two bodies through all of time, past and future.

However, the universe doesn't contain only two bodies. It contains uncounted trillions. The next step then is to build up towards taking them all into account by solving the 'three-body problem'. Given three bodies in the universe, with known position and motion, what will their relative positions be at all given times?

And right there, astronomers are stymied. There is no general solution to such a problem. For that reason there is no use going on to the 'octillion-body problem' represented by the actual universe.

Fortunately, astronomers are not halted in a practical sense. The theory may be lacking but they can get along. Suppose, for instance, that it was necessary to calculate the orbit of the earth about the sun so that the relative positions could be calculated for the next million years. If the sun and the earth were all that existed, the problem would be a trivial one to solve. But the gravity of the moon must be considered, and of Mars and the other planets, and, for complete exactness, even the stars.

Fortunately, the sun is so much bigger than any other body in the vicinity and so much closer than any other really massive body, that its gravitational field drowns out all others. The orbit obtained for the earth by calculating a simple two-body situation is almost right. You then calculate the minor effect of the closer bodies and make corrections. The closer you want to pinpoint the exact orbit, the more corrections you must make, covering smaller and smaller perturbations.

The principle is clear but the practice can become tedious, to be sure. The equation that gives the motion of the moon with reasonable exactness covers many hundreds of pages. But that is good enough to predict the time and position of eclipses with great correctness for long periods of time into the future.

Nevertheless, astronomers are not satisfied. It is all very well to work out orbits on the basis of successive approximations, but how beautiful and elegant it would be to prepare an equation that would interrelate all bodies in a simple and grand way. Or three bodies, anyway.

The man who most closely approached this ideal was the French astronomer Joseph Louis Lagrange. In 1772, he actually found certain very specialised cases in which the three-body problem could be solved.

Imagine two bodies in space with the mass of body A at least 25·8 times that of body B, so that B can be said to revolve about a virtually motionless A, as Jupiter, for instance, revolves about the sun. Next imagine a third body, C, of comparatively insignificant mass, so that it does not disturb the gravitational relationship of A and B. Lagrange found that it was possible to place body C at certain points in relationship to bodies A and B, so that C would revolve about A in perfect step with B. In that way the relative positions of all three bodies would be known for all times.

There are five points at which body C can be placed; and they are, naturally enough, called 'Lagrangian points'. Three of them, L_1, L_2 and L_3 are on the line connecting A and B. The first point, L_1, places small body C between A and B. Both L_2 and L_3 lie on the line also, but on the other side of A in the first case and on the other side of B in the next.

These three Lagrangian points are not important. If any body located at one of those points, moves ever so slightly off position due to the perturbation of some body outside the system, the resulting effect of the gravitational fields of A and B is to throw C still farther off the point. It is like a long stick balanced on edge. Once that stick tips ever so slightly, it tips more and more and falls.

However, the final Lagrangian points are not on the line connecting bodies A and B. Instead, they form equilateral triangles with A and B. As B revolves about A, L_4 is the point that moves before B at a constant angle of 60°, while L_5 moves behind it at a constant 60°.

These last two points are stable. If an object at either point moves slightly off position, through outside perturbations, the effect of the gravitational fields of A and B is to bring them back. In this way, objects at L_4 and L_5 oscillate about the true Lagrangian point, like a long stick balanced at the end of a finger which adjusts its position constantly to prevent falling.

Of course, if the stick tips *too* far out of vertical it will fall despite the balancing efforts of the finger. And if a body moves *too* far away from the Lagrangian point it will be lost.

At the time Lagrange worked this out, no examples of objects located at Lagrangian points were known anywhere in the universe. However, in 1906, a German astronomer, Max Wolf, discovered an asteroid, which he named Achilles after the Greek hero of the *Iliad*. It was unusually far out for an asteroid. In fact, it was as far from the sun as Jupiter was.

An analysis of its orbit showed that it always remained near the Lagrangian point, L_4, of the sun-Jupiter system. Thus, it stayed a fairly constant 480,000,000 miles ahead of Jupiter in its motion about the sun.

Some years later, another asteroid was discovered in the L_5 position of the sun-Jupiter system, and was named Patroclus, after Achilles' beloved friend. It moves about the sun in a position that is a fairly constant 480,000,000 miles behind Jupiter.

Other asteroids were in time located at both points; at the present time, fifteen of these asteroids are known: ten in L_4 and five in L_5. Following the precedent of Achilles, all have been named for characters in the *Iliad*. And since the *Iliad* deals with the Trojan War, all the bodies in both positions are lumped together as the 'Trojan asteroids'.

Since the asteroids at position L_4 include Agamemnon, the Greek leader, they are sometimes distinguished as the 'Greek group'. The asteroids at position L_5 include the one named for the Trojan king Priamus (usually known as 'Priam' in English versions of the *Iliad*), and are referred to as the 'pure Trojan group'.

It would be neat and tidy if the Greek group contained only Greeks and the pure Trojan group only Trojans. Unfortunately, this was not thought of. The result is that the Trojan hero Hector is part of the Greek group and the Greek hero Patroclus is part of the pure Trojan group. It is a situation that would strike any classicist with apoplexy. It makes even myself feel a little uncomfortable, and I am only the very mildest of classicists indeed.

The Trojan asteroids remain the only known examples of objects at Lagrangian points. They are so well known, however, that L_4 and L_5 are commonly known as 'Trojan positions'.

External perturbing forces, particularly that of the planet Saturn, keep the asteroids oscillating about the central points. Sometimes the oscillations are wide; a particular asteroid may be as much as 100,000,000 miles from the Lagrangian point.

Eventually, a particular asteroid may be pulled too far outward, and would then adopt a non-Trojan orbit. On the other hand, some asteroid now independent, may happen to be perturbed into a spot close to the Lagrangian points and be trapped. In the long run, the Trojan asteroids may change identities, but there will always be some there.

Undoubtedly, there are many more than fifteen Trojan asteroids. Their distance from us is so great that only fairly large asteroids, close to one hundred miles in diameter, can be seen. Still, there are certainly dozens and even hundreds of smaller chunks, invisible to us, that chase Jupiter or are chased in an eternal race that nobody wins.

There must be many Trojan situations in the universe. I wouldn't be surprised if every pair of associated bodies which met the 25·8 to 1 mass-ratio requirement was accompanied by rubble of some sort at the Trojan positions.

Knowing that the rubble exists doesn't mean that it can be spotted, however; certainly it can be detected nowhere outside the solar system. Three related stars could be spotted, of course, but for a true Trojan situation, one body must be of insignificant mass, and it could not be seen by any technique now at our disposal.

Within the solar system, by far the largest pair of bodies are the sun and Jupiter. The bodies trapped at the Lagrangian points of that system could be fairly large and yet remain negligible in mass in comparison to Jupiter.

The situation with respect to Saturn would be far less favourable. Since Saturn is smaller than Jupiter, the asteroids at the Trojan position associated with Saturn would be smaller on the average. They would be twice as far from us as those of Jupiter are, so that they would also be dimmer. They would thus be very difficult to see; and the fact of the matter is that no Saturnian Trojans have been found. The case is even worse for Uranus, Neptune and Pluto.

As for the small inner planets, there any rubble in the

Trojan position must consist of small objects indeed. That alone would make them nearly impossible to see, even if they existed. In addition, particularly in the case of Venus and Mercury, they would be lost in the glare of the sun.

In fact, astronomers do not really expect to find the equivalent of Trojan asteroids for any planet of the solar system other than Jupiter, until such time as an astronomical laboratory is set up outside the earth or, better yet, until spaceships actually explore the various Lagrangian points.

Yet there is one exception to this, one place where observation from the earth's surface can turn up something and, in fact, may have done so. That is a Lagrangian point that is not associated with a sun-planet system, but with a planet-satellite system. Undoubtedly you are ahead of me and know that I am referring to the earth and the moon.

The fact that the earth has a single satellite was known as soon as man grew intelligent enough to become a purposeful observer. Modern man with all his instruments has never been able to find a second one.[4] Not a natural one, anyway. In fact, astronomers are quite certain that, other than the moon itself, no body that is more than, say, half a mile in diameter, revolves about the earth.

This does not preclude the presence of any number of very small particles. Data brought back by artificial satellites would seem to indicate that the earth is surrounded by a ring of dust particles something after the fashion of Saturn, though on a much more tenuous scale.

Visual observation could not detect such a ring except in places where the particles might be concentrated in unusually high densities. The only spots where the concentration could be great enough would be at the Lagrangian points, L_4 and L_5, of the earth-moon system (Since the earth is more than 25.8 times as massive as the moon — it is

81 times as massive in point of fact — objects at those points would occupy a stable position.)

Sure enough, in 1961, a Polish astronomer, K. Kordylewski, reported actually spotting two very faintly luminous patches in these positions. Presumably, they represent dust clouds trapped there.

And in connection with these 'cloud satellites', I have thought up a practical application of Lagrangian points which, as far as I know, is original.

As we all know, one of the great problems brought upon us by the technology of the space age is that of the disposal of radioactive wastes. Many solutions have been tried or have been suggested. The wastes are sealed in strong containers or, as is suggested, fused in glass. They may be buried underground, stored in salt-mines or dropped into an abyss.

No solution that leaves the radioactivity upon the earth, however, is wholly satisfactory; so some bold souls have suggested that eventually measures will be taken to fire the wastes into space.

The safest procedure one can possibly imagine is to shoot these wastes into the sun. This, however, is not an easy thing to do at all. It would take less energy to shoot them to the moon, but I'm sure that astronomers would veto that. It would be still easier simply to shoot them into an orbit about the sun, and easiest of all to shoot them into an orbit about the earth.

In either of these latter cases, however, we run the risk, in the long run, of cluttering up the inner portions of the solar system, particularly the neighbourhood of the earth, with gobs of radioactive material. We would be living in the midst of our own refuse, so to speak.

Granted that space is large and the amount of refuse, in comparison, is small, so that collisions or near-collisions between spaceships and radioactive debris would be highly improbable, it could still lead to trouble in the long run.

Consider the analogy of our atmosphere. All through history, man has freely poured gaseous wastes and smoky particles into it in the certainty that all would be diluted far past harm; yet air pollution has now become a major problem. Well, let's not pollute space.

One way out is to concentrate our wastes into one small portion of space and make sure it stays there. Those regions of space can then be marked off-limits and everything else will be free of trouble.

To do this, one would have to fire the wastes to one or the other of the Trojan positions associated with the earth-moon system in such a way as to leave it trapped there. Properly done, the wastes would remain at those points, a quarter-million miles from the moon and a quarter-million miles from the earth, for indefinite periods, certainly long enough for the radiation to die down to non-dangerous levels.

Naturally, the areas would be a death trap for any ship passing through — a kind of 'Trojan hearse', in fact. Still, it would be a small price to pay for solving the radioactive ash disposal problem, just as this pun is a small price to pay for giving me a title for the chapter.

CHAPTER FIVE

BY JOVE!

Suppose we ask ourselves a question: On what world of the solar system (other than earth itself, of course) are we most likely to discover life?

I imagine I can plainly hear the unanimous answering shout, '*Mars!*'

The argument goes, and I know it by heart, because I have used it myself a number of times, that Mars may be a little small and a little cold and a little short on air, but it isn't too small, too cold, or too airless to support the equivalent of primitive plant life. On the other hand, Venus and Mercury are definitely too hot, the moon is airless, and the remaining satellites of the solar system, and the planetoids as well (to say nothing of Pluto), are too cold, too small, or both.

And then we include a phrase which may go like this: 'As for Jupiter, Saturn, Uranus and Neptune, we can leave them out of consideration altogether.'

However, Carl Sagan, an astronomer at Cornell University, doesn't take this attitude at all, and a paper of his on the subject has lured me into doing a bit of thinking on the subject of the outer planets.

Before Galileo's time, there was nothing to distinguish Jupiter and Saturn (Uranus and Neptune not having yet been discovered) from the other planets, except for the fact that they moved more slowly against the starry background than did the other planets and were, therefore, presumably farther from the earth.

The telescope, however, showed Jupiter and Saturn as discs with angular widths that could be measured. When the distances of the planets were determined, those angular widths could be converted into miles, and the result was a shocker. As compared with an earthly equatorial diameter of 7,950 miles, Jupiter's diameter across its equator was 88,800, while Saturn's was 75,100.

The outer planets were giants!

The discovery of Uranus in 1781 and Neptune in 1846 added two more not-quite-so-giants, for the equatorial diameter of Uranus is 31,000 miles and that of Neptune, at latest measurement, is about 28,000 miles.

The disparity in size between these planets and our own tight little world is even greater if one considers volume, because that varies as the cube of the diameter. In other words, if the diameter of Body A is ten times the diameter of Body B, then the volume of Body A is ten times ten times ten, or a thousand times the volume of Body B. Thus, if we set the volume of the earth equal to 1, the volumes of the giants are given in Table 7.

Table 7

PLANET	VOLUME (EARTH = 1)
Jupiter	1,300
Saturn	750
Uranus	60
Neptune	40

Each of the giants has satellites. It is easy to determine the distance of the various satellites from the centre of the primary planet by measuring the angular separation. It is also easy to time the period of revolution of the satellite. From those two pieces of datum, one can quickly obtain the mass of the primary (*see* Chapter 3).

In terms of mass, the giants remain giants, naturally. If the mass of earth is taken as 1, the masses of the giants are as given in Table 8.

Table 8

PLANET	MASS (EARTH = 1)
Jupiter	318
Saturn	95
Uranus	15
Neptune	17

The four giants contain virtually all the planetary mass of the solar system, Jupiter alone possessing about 70 per cent of the total. The remaining planets, plus all the satellites, planetoids, comets and, for that matter, meteoroids, contain well under 1 per cent of the total planetary mass. Outside intelligences, exploring the solar system with true impartiality, would be quite likely to enter the sun in their records thus: Star X, spectral class GO, 4 planets plus debris.

But take another look at the figures on mass. Compare them with those on volume and you will see that the mass is consistently low. In other words, Jupiter takes up 1,300 times as much room as the earth does, but contains only 318 times as much matter. The matter in Jupiter must therefore be spread out more loosely, which means, in more formal language, that Jupiter's density is less than that of the earth.

If we set the earth's density equal to 1, then we can obtain the densities of the giants by just dividing the figure for the relative mass by the figure for the relative volume. The densities of the giants are given in Table 9.

On this same scale of densities, the density of water is 0·182. As you see, then, Neptune, the densest of the giants, is only about 2¼ times as dense as water, while Jupiter and Uranus are only 1½ times as dense, and Saturn is actually less dense than water.

Table 9

PLANET	DENSITY (EARTH = 1)
Jupiter	0·280
Saturn	0·125
Uranus	0·250
Neptune	0·425

I remember seeing an astronomy book that dramatised this last fact by stating that if one could find an ocean large enough, Saturn would float in it, less than three-fourths submerged. And there was a very impressive illustration, showing Saturn, rings and all, floating in a choppy sea.

But don't misinterpret this matter of density. The first thought anyone naturally might have is that because Saturn's overall density is less than that of water, it must be made of some corklike material. This, however, is not so, as I can explain easily.

Jupiter has a striped or banded appearance, and certain features upon its visible surface move around the planet at a steady rate. By following those features, the period of rotation can be determined with a high degree of precision; it turns out to be 9 hours, 50 minutes, and 30 seconds. (With increasing difficulty, the period of rotation can be determined for the more distant giants as well.)

But here a surprising fact is to be noted. The period of rotation I have given is that of Jupiter's equatorial surface.

Other portions of the surface rotate a bit more slowly. In fact, Jupiter's period of rotation increases steadily as the poles are approached. This alone indicates we are not looking at a solid surface, for that would have to rotate all in one piece.

The conclusion is quite clear. What we see as the surface of Jupiter, and of the other giants, are the clouds of its atmosphere. Beneath those clouds must be a great depth of atmosphere, far denser than our own, and yet far less dense than rock and metal. It is because the atmosphere of the giant planets is counted in with their volume that their density appears so low. If we took into account only the core of the planet, underlying the atmosphere, we could find a density as great as that of earth's, or, most likely, greater.

But how deep is the atmosphere?

Consider that, fundamentally, the giant planets differ from the earth chiefly in that, being further from the sun and therefore colder through their history, they retain a much larger quantity of the light elements — hydrogen, helium, carbon, nitrogen and oxygen. Helium forms no compounds but remains as a gas. Hydrogen is present in large excess so it remains as a gas, too, but it also forms compounds with carbon, nitrogen and oxygen, to form methane, ammonia and water, respectively. Methane is a gas, and, at the earth's temperature, so is ammonia, but water is a liquid. If the earth's temperature were to drop to −100° C. or below, both ammonia and water would be solid, but methane would still be a gas.

As a matter of fact, all this is not merely guesswork. Spectroscopic evidence does indeed show that Jupiter's atmosphere is hydrogen and helium in a three-to-one ratio with liberal admixtures of ammonia and methane. (Water is not detected, but that may be assumed to be frozen out.)

Now, the structure of the earth can be portrayed as a central solid body of rock and metal (the lithosphere), surrounded by a layer of water (the hydrosphere), which is in turn surrounded by a layer of gas (the atmosphere).

The light elements in which the giant planets are particularly rich would add to the atmosphere and to the hydrosphere, but not so much to the lithosphere. The picture would therefore be of a central lithosphere larger than that of the earth, but not necessarily enormously larger, surrounded by a gigantic hydrosphere and an equally gigantic atmosphere.

But how gigantic is gigantic?

Here we can take into consideration the polar flattening of the giants. Thus, although Jupiter is 88,800 miles in diameter along the equator, it is only 82,800 miles in diameter from pole to pole. This is a flattening of 7 per cent, compared to a flattening of about ·33 per cent for the earth. Jupiter has a visibly elliptical appearance for that reason. Saturn's aspect is even more extreme, for its equatorial diameter is 75,100 miles while its polar diameter is 66,200 miles, a flattening of nearly 12 per cent. (Uranus and Neptune are less flattened than are the two larger giants.)

The amount of flattening depends partly on the velocity of rotation and the centrifugal effect which is set up. Jupiter and Saturn, although far larger than the earth, have periods of rotation of about 10 hours as compared with our own 24. Thus, the Jovian surface, at its equator, is moving at a rate of 25,000 miles an hour, while the earth's equatorial surface moves only at a rate of 1,000 miles an hour. Naturally, Jupiter's surface is thrown farther outward than earth's is (even against Jupiter's greater gravity), so that the giant planet bulges more at the equator and is more flattened at the poles.

However, Saturn is distinctly smaller than Jupiter and has

a period of rotation some twenty minutes longer than that of Jupiter. It exerts a smaller centrifugal effect at the equator; and even allowing for its smaller gravity, it should be less flattened at the poles than Jupiter is. However, it is more flattened. The reason for this is that the degree of flattening depends also on the distribution of density, and if Saturn's atmosphere is markedly thicker than Jupiter's, flattening will be greater.

The astronomer Rupert Wildt has estimated what the size of the lithosphere, hydrosphere and atmosphere[1] would have to be on each planet in order to give it the overall density it was observed to have plus its polar flattening. (This picture is not accepted by astronomers generally, but let's work with it anyway.[2]) The figures I have seen are included in Table 10, to which I add figures for the earth as a comparison:

Table 10

	LITHOSPHERE (RADIUS IN MILES)	HYDROSPHERE (THICKNESS IN MILES)	ATMOSPHERE (THICKNESS IN MILES)
Jupiter	18,500	17,000	8,000
Saturn	14,000	8,000	16,000
Uranus	7,000	6,000	3,000
Neptune	6,000	6,000	2,000
Earth	3,975	2	8[3]

As you see, Saturn, though smaller than Jupiter, is pictured as having a much thicker atmosphere, which accounts for its low overall density and its unusual degree of flattening. Neptune has the shallowest atmosphere and is therefore the densest of the giant planets.

Furthermore, you can see that the earth isn't *too* pygmyish in comparison with the giants, if the lithosphere alone is considered. If we assume that the lithospheres are all of equal density and set the mass of earth's lithosphere equal to 1, then the masses of the others are shown in Table 11.

Table 11

PLANET	MASS OF LITHOSPHERE (EARTH = 1)
Jupiter	100
Saturn	45
Uranus	$5\frac{1}{2}$
Neptune	$3\frac{1}{2}$

It is the disparity of the hydrosphere and atmosphere that blows up the giants to so large a size.

To emphasise this last fact, it would be better to give the size of the various components in terms of volume rather than of thickness. In Table 12 the volumes are therefore given in trillions of cubic miles. Once again, the earth is included for purposes of comparison:

Table 12

	LITHOSPHERE VOLUME	HYDROSPHERE VOLUME	ATMOSPHERE VOLUME
Jupiter	27	161	155
Saturn	11·5	33	185
Uranus	1·4	7·8	8·4
Neptune	0·9	6·4	4·2
Earth	0·26	0·00033	0·0011

As you can see at a glance, the lithosphere of the giant planets makes up only a small part of the total volume, whereas it makes up almost all the volume of the earth. This shows up more plainly if, in Table 13, we set up the volume of each component as a percentage of its planet's total volume. Thus:

Table 13

	LITHOSPHERE (% OF PLANET'S VOLUME)	HYDROSPHERE (% OF PLANET'S VOLUME)	ATMOSPHERE (% OF PLANET'S VOLUME)
Jupiter	7·7	47·0	45·3
Saturn	4·8	14·4	80·8
Uranus	8·0	44·3	47·7
Neptune	8·0	55·5	36·5
Earth	99·45	0·125	0·425

The difference can't be made plainer. Whereas the earth is about 99·5 per cent lithosphere, the giant planets are only 8 per cent, or less, lithosphere. About one-third of Neptune's apparent volume is gas. In the case of Jupiter and Uranus, the gas volume is one-half the total, and in the case of Saturn, the least dense of the four, the gas volume is fully four-fifths of the total. The giant planets are sometimes called the 'gas giants' and, as you see, that is a good name, particularly for Saturn.

This is a completely alien picture we have drawn of the giant planets. The atmospheres are violently poisonous, extremely deep and completely opaque, so that the surface of the planet is entirely and permanently dark even on the 'sunlit side'. The atmospheric pressure is gigantic; and from what we can see of the planets, the atmosphere seems to be beaten into the turmoil of huge storms.

The temperatures of the planets are usually estimated as ranging from a −100° C. maximum for Jupiter to a −230°

C. minimum for Neptune, so that even if we could survive
the buffeting and the pressures and the poisons of the
atmosphere, we would land on a gigantic, planet-covering,
thousands-of-miles-thick layer of ammoniated ice.

Not only is it inconceivable for man to land and live on
such a planet, but it seems inconceivable that any life at all
that even remotely resembles our own could live there.

Are there any loopholes in this picture?

Yes, a very big one, possibly, and that is the question of
the temperature. Jupiter may not be nearly as cold as we
have thought.

To be sure, it is about five times as far from the sun as we
are, so that it receives only one twenty-fifth as much solar
radiation. However, the crucial point is not how much radiation it receives but how much it keeps. Of the light it
receives from the sun, four-ninths is reflected and the remaining five-ninths is absorbed. The absorbed portion does
not penetrate to the planetary surface as light, but it gets
there just the same — as heat.

The planet would ordinarily reradiate this heat as long-
wave infrared, but the components of Jupiter's atmosphere,
notably the ammonia and methane, are quite opaque to
infrared, which is therefore retained forcing the temperature
to rise. It is only when the temperature is quite high that
enough infrared can force its way out of the atmosphere to
establish a temperature equilibrium.

It is even possible that the surface temperature of Jupiter,
thanks to this 'greenhouse effect', is as high as that of earth.

The other giant planets may also have temperatures
higher than those usually estimated, but the final equilibrium
would very likely be lower than that of Jupiter's since the
other planets are further from the sun. Perhaps Jupiter is the
only giant planet with a surface temperature above 0° C.

This means that Jupiter, of all the giant planets, could be

the only one with a liquid hydrosphere. Jupiter would have a vast ocean, covering the entire planet (by the Wildt scheme) and 17,000 miles deep.[4]

On the other hand, Venus also has an atmosphere that exerts a greenhouse effect, raising its surface temperature to a higher level than had been supposed. Radio-wave emission from Venus indicates its surface temperature to be much higher than the boiling point of water, so that the surface of Venus is powdery dry with all its water supply in the cloud layer overhead.

A strange picture. The planetary ocean that has been so time-honoured a science-fictional picture of Venus has been pinned to the wrong planet all along. It is Jupiter that has the world-wide ocean, by Jove!

Considering the Jovian ocean, Professor Sagan (to whom I referred at the beginning of this chapter) says: 'At the present writing, the possibility of life on Jupiter seems somewhat better than the possibility of life on Venus.'

This is a commendably cautious statement, and as far as a scientist can be expected to go in a learned journal. However, I, myself, on this particular soapbox, don't have to be cautious at all, so I can afford to be much more sanguine about the Jovian ocean. Let's consider it for a moment.

If we accept Wildt's picture, it is a big ocean, nearly 500,000 times as large as earth's ocean and, in fact, 620 times as voluminous as all the earth. This ocean is under the same type of atmosphere that, according to current belief, surrounded the earth all the time life developed on our planet. All the simple compounds — methane, ammonia, water, dissolved salts — would be present in unbelievable plenty by earthly standards.

Some source of energy is required for the building up of these organic compounds, and the most obvious one is the ultraviolet radiation of the sun. The quantity of ultraviolet rays that reaches Jupiter is, as aforesaid, only one twenty-fifth that which reaches the earth, and none of it can get very far into the thick atmosphere.

Nevertheless, the ultraviolet rays must have some effect, because the coloured bands in the Jovian atmosphere are very likely to consist of free radicals (that is, energetic molecular fragments) produced out of ordinary molecules by the ultraviolet.

The constant writhing of the atmosphere would carry the free radicals downward where they could transfer their energy by reacting with simple molecules to build up complex ones.

Even if ultraviolet light is discounted as an energy source, two other sources remain. There is first, lightning. Lightning in the thick soup that is called a Jovian atmosphere may be far more energetic and continuous than it ever is or was on earth. Secondly, there is always natural radioactivity.

Well, then, why can't the Jovian ocean breed life? The temperature is right. The raw material is there. The energy supply is present. All the requirements that were sufficient to produce life in earth's primordial ocean are present also on Jupiter (if the picture drawn in this article is correct), only more and better.

One might wonder whether life could withstand the Jovian atmospheric pressures and storms, to say nothing of the Jovian gravity. But the storms, however violent, could only roil up the outer skin of a 17,000-mile-deep ocean. A few hundred feet below the surface, or a mile below, if you like, there would be nothing but the slow ocean currents.

As for gravity, forget it. Life within the ocean can ignore gravity altogether, for buoyancy neutralises its effects, or almost neutralises it.

No, none of the objections stand up. To be sure, life must

originate and develop on Jupiter in the absence of gaseous oxygen, but that is exactly one of the conditions under which life originated and developed on earth. There are living creatures on earth right now that can live without oxygen.

So once again let's ask the question: On what world of the solar system (other than earth itself, of course) are we most likely to discover life?

And now, it seems to me, the answer must be: On Jupiter, by Jove!

Of course, life on Jupiter would be pitifully isolated. It would have a vast ocean to live in, but the far, far vaster outside universe would be closed forever to them.

Even if some forms of Jovian life developed an intelligence comparable to our own (and there are reasonable arguments to suggest that true sea life — and before you bring up the point, dolphins are descendants of land-living creatures — would not develop such intelligence), they could do nothing to break the isolation.

It is highly unlikely that even a manlike intelligence could devise methods that would carry itself out of the ocean, through thousands of miles of violent, souplike atmosphere, against Jupiter's colossal gravity, in order to reach Jupiter's inmost satellite and, from its alien surface, observe the universe.

And as long as life remained in the Jovian ocean, it would receive no indication of an outside universe, except for a non-directed flow of heat, and excessively feeble microwave radiation from the sun and a few other spots. Considering the lack of supporting information, the microwaves would be as indecipherable a phenomenon as one could imagine, even if it were sensed.

But let's not be sad; let's end on a cheerful note.

If the Jovian ocean is as rich in life as our own is, then

$1/70,000$ of its mass would be living matter. In other words, the total mass of sea life on Jupiter would be one-eighth the mass of our moon, and that's a lot of mass for a mess of fish.

What fishing-grounds Jupiter would make if it could be reached somehow!

And, in view of our population explosion, just one question to ponder over ... Do you suppose that Jovian life might be edible?

CHAPTER SIX

SUPERFICIALLY SPEAKING

For the last century, serious science-fiction writers, from Edgar Allan Poe onward, have been trying to reach the moon; and now governments are trying to get into the act.[1] It kills some of the romance of the deal to have the project become a 'space spectacular' designed to show up the other side, but if that's what it takes to get there, I suppose we can only sigh and push on.

So far, however, governments are only interested in *reaching* the moon, and as science-fiction fans we ought to remain one step ahead of them and keep our eyes firmly fixed on *populating* the moon. Naturally, we can ignore such little problems as the fact that air and water are missing on the moon. Perhaps we can bake water out of the deep-lying rock and figure out ways of chipping oxygen out of silicates.[2] We can live underground to get away from the heat of the day and the cold of the night.

In fact, with the sun shining powerfully down from a cloudless sky for two weeks at a time, solar batteries might be able to supply moon colonists with tremendous quantities of energy.

Maybe the land with the high standard of living of the future will be up there in the sky. Etched into some of the craters, perhaps, large enough to be clearly seen through a small telescope, could be a message that starts: 'Send me your tired, your poor, your huddled masses yearning to breathe free . . .'

Who knows?

But if the moon is ever to be a second earth and is to siphon off some of our population, there is a certain significant statistic about it that we ought to know. That is, its size.

The first question is, what do we mean by 'size'?

The size of the moon is most often given in terms of its diameter, because once the moon's distance has been determined, its diameter can be obtained by direct measurement.

Since the moon's diameter is 2,160 miles and the earth's is 7,914 miles, most people cannot resist the temptation of saying that the moon is one-quarter the size of the earth, or that the earth is four times the size of the moon. (The exact figure is that the moon is 0·273 times the size of the earth, from this viewpoint, or that the earth is 3·66 times the size of the moon.)

All this makes the moon appear quite a respectably-sized world.

But, let's consider size from a different standpoint. Next to diameter, the most interesting statistic about a body of the solar system is its mass, for upon that depends the gravitational force it can exert.

Now, mass varies as the cube of the diameter, all things being equal. If the earth is 3·66 times the size of the moon, diameterwise, it could be 3·66×3·66×3·66 or 49 times the size of the moon, masswise. (Hmm, there's something to be said for this Madison Avenue speech monstrosity, conveniencewise.) But that is only if the densities of the two bodies being compared are the same.

As it happens, the earth is 1·67 times as dense as the moon, so that the discrepancy in mass is even greater than a simple cubing would indicate. Actually, the earth is 81 times as massive as the moon.

This is distressing because now, suddenly, the moon has grown a bit pygmyish on us, and the question arises as to which we ought really to say. Is the moon one-quarter the size of the earth or is it only $1/_{81}$ the size of the earth?

Actually, we ought to use whichever comparison is meaningful under a particular set of circumstances, and as far as populating the moon is concerned, neither is directly meaningful. What counts is the surface area, the *superficial* size of the moon.

On any sizable world, under ordinary circumstances, human beings will live on the surface. Even if they dig underground to escape an unpleasant environment, they will do so only very slightly, when compared with the total diameter, on any world the size of the earth or even that of the moon.

Therefore, the question that ought to agitate us with respect to the size of the moon is: What is its surface area in comparison with that of the earth? In other words, what is its size, superficially speaking?

This is easy to calculate because surface area varies as the square of the diameter. Here density has no effect and need not be considered. If the earth has a diameter 3·66 times that of the moon, it has a surface area 3·66 × 3·66 or 13·45 times that of the moon.

But this doesn't satisfy me. The picture of a surface that is equal to $1/_{13.45}$ that of the earth isn't dramatic enough. What does it mean exactly? Just how large is such a surface?

I've thought of an alternate way of dramatising the moon's surface, and that of other areas, and it depends on the fact that a good many Americans these days have been jetting freely about the United States. This gives them a good conceptual feeling of what the area of the United States is like, and we can use that as a unit. The area of all fifty states is 3,628,000 square miles and we can call that 1 USA unit.

To see how this works, look at Table 14, which includes a sampling of geographic divisions of our planet with their areas given in USAs.

Table 14

GEOGRAPHIC DIVISION	AREA (IN USAS)	GEOGRAPHIC DIVISION	AREA (IN USAS)
Australia	0·82	North America	54·30
Brazil	0·91	Africa	2·50
Canada	0·95	Asia	3·20
United States	1·00	Indian Ocean	4·70
Europe	1·07	Atlantic Ocean	7·80
China	1·19	Total land surface	8·80
Arctic Ocean	1·50	Pacific Ocean	17·50
Antarctica	1·65	Total water surface	17·60
South America	1·90	Total surface	36·80
Soviet Union	2·32		

Now, you see, when I say that the moon's surface is 4·03 USAs, you know at once that the colonization of the moon will make available to humanity an area of land equal to four times that of the United States or 1·75 times that of the Soviet Union. To put it still another way, the area of the moon is just about halfway between that of Africa and Asia.

But let's go further and assume that mankind is going to colonise all the solar system that it can colonise or that is worth colonising. When we say 'can colonise', we eliminate, at least for the foreseeable future, the 'gas giants', that is Jupiter, Saturn, Uranus and Neptune. (For some comments on them, however, see the previous chapter.)

That still leaves four planets: Mercury, Venus, Mars, and

(just to be complete — and extreme) Pluto. In addition, there are a number of sizable satellites, aside from our own moon, that are large enough to seem worth colonising. These include the four large satellites of Jupiter (Io, Europa, Ganymede and Callisto), the two large satellites of Saturn (Titan and Rhea), and Neptune's large satellite (Triton).

The surface areas of these bodies are easily calculated; the results are given in Table 15, with the earth and moon included for comparison. As you can see, if we exclude the sun and the gas giants, there are a round dozen bodies in the solar system with a surface area in excess of 1 USA, and a thirteenth with an area just short of that figure.

Table 15

PLANET OR SATELLITE   SURFACE AREA (USAS)

Earth	54·3
Pluto	(54)??
Venus	49·6
Mars	15·4
Callisto	9·0
Ganymede	8·85
Mercury	8·30
Titan	7·30
Triton	6·80
Io	4·65
Moon	4·03
Europa	3·30
Rhea	0·86

The total surface area available on this baker's dozen of worlds is roughly equal to 225 USAs. Of this, the earth itself represents fully one quarter, and the earth is already colonised, so to speak, by mankind. Another quarter is represented by Pluto, the colonisation of which, with the best will in the world, must be considered as rather far off.[3]

Of what is left (about 118 USAs), Venus, Mars and the moon make up some five-ninths. Since these represent the worlds that are closest and therefore the most easily reached and colonised, there may be quite a pause before humanity dares the sun's neighbourhood to reach Mercury, or sweeps outward to the large outer satellites. It might seem that the extra pickings are too slim.

However, there are other alternatives, as I shall explain.

So far, I have not considered objects of the solar system that are less than 1,000 miles in diameter (which is the diameter of Rhea). At first glance, these might be considered as falling under the heading of 'not worth colonising' simply because of the small quantity of surface area they might be expected to contribute. In addition, gravity would be so small as to give rise to physiological and technological difficulties, perhaps.

However, let's ignore the gravitational objection, and concentrate on the surface area instead.

Are we correct in assuming that the surface area of the minor bodies is small enough to ignore? There are, after all, twenty-three satellites in the solar system with diameters of less than 1,000 miles, and that's a respectable number. On the other hand, some of these satellites are quite small. Deimos, the smaller satellite of Mars, has a diameter that isn't more than 7·5 miles.[4]

To handle the areas of smaller worlds, let's make use of another unit. The American city of Los Angeles covers 450 square miles. We can set that equal to 1 LA. This is convenient because it means there are just about 8,000 LAs in 1 USA.

A comparison of the surface areas of the minor satellites of the solar system is presented in Table 16. (I'll have to point out that the diameters of all these satellites are quite uncertain and that the surface areas as given are equally uncertain. However, they are based on the best information available to me.)

Table 16

SATELLITE (PRIMARY)	SURFACE AREA (LAS)
Iapetus (Saturn)	4,450
Tethys (Saturn)	3,400
Dione (Saturn)	3,400
Titania (Uranus)	2,500
Oberon (Uranus)	2,500
Mimas (Saturn)	630
Enceladus (Saturn)	630
Ariel (Uranus)	630
Umbriel (Uranus)	440
Hyperion (Saturn)	280
Phoebe (Saturn)	280
Nereid (Neptune)	120
Amalthea (Jupiter)	70
Miranda (Uranus)	45
VI (Jupiter)	35
VII (Jupiter)	6·5
VIII (Jupiter)	6·5
IX (Jupiter)	1·5
XI (Jupiter)	1·5
XII (Jupiter)	1·5
Phobos (Mars)	1·5
X (Jupiter)	0·7
Deimos (Mars)	0·4

The total area of the minor satellites of the solar system thus comes to just under 20,000 LAs or, dividing by 8,000, to about 2·5 USAs. All twenty-three worlds put together have little more than half the surface area of the moon, or, to put it another way, have just about the area of North America.

This would seem to confirm the notion that the minor satellites are not worth bothering about, but . . . let's think again. All these satellites, lumped together, have just a trifle over one-sixth the volume of the moon, and yet they have more than half its surface area.

This should remind us that the smaller a body, the larger its surface area in proportion to its volume. The surface area of any sphere is equal to $4\pi r^2$, where r is its radius. This means that the earth, with a radius of roughly 4,000 miles, has a surface area of roughly 200,000,000 square miles.

But suppose the material of the earth is used to make up a series of smaller worlds each with half the radius of the earth. Volume varies as the cube of the radius, so the material of the earth can make up no less than eight 'half-earths', each with a radius of roughly 2,000 miles. The surface area of each 'half-earth' would be roughly 50,000,000 square miles, or twice the area of the original earth.

If we consider a fixed volume of matter, then, the smaller the bodies into which it is broken up, the larger the total surface area it exposes.

You may feel this analysis accomplishes nothing, since the twenty-three minor satellites do not, in any case, have much area. Small though they are, the total area comes to that of North America and no more.

Ah, but we are not through. There are still the minor planets, or asteroids.

It is estimated that all the asteroids put together have a mass about 1 per cent that of the earth. If all of them were somehow combined into a single sphere, with an average density equal to that of the earth, the radius of that sphere would be 860 miles and the diameter, naturally, 1,720 miles. It would be almost the size of one of Jupiter's satellites,

Europa, and its surface area would be 2·6 USAs, or just about equal to that of all the minor satellites put together.

But the asteroids do not exist as this single fictitious sphere but as a large number of smaller pieces, and here is where the increase in surface area comes in. The total number of asteroids is estimated to be as high as 100,000; and if that figure is correct, then the average asteroid has a diameter of 35 miles, and the total surface area of all 100,000 would then come to as much as 130 USAs.

This means that the total surface area of the asteroids is equal to slightly more than that of the earth, Venus, Mars, and the moon all lumped together. It is 7·5 times the area of the earth's land surface. Here is an unexpected bonanza.

Furthermore, we can go beyond that. Why restrict ourselves only to the surface of the worlds? Surely we can dig into them and make use of the interior materials otherwise beyond our reach. On large worlds, with their powerful gravitational forces, only the outermost skin can be penetrated, and the true interior seems far beyond our reach. On an asteroid, however, gravity is virtually nil and it would be comparatively easy to hollow it out altogether.

I made use of this notion in a story I once wrote[5] which was set on a fictitious asteroid called Elsevere. A visitor from earth is being lectured by one of the natives, as follows:

'We are not a small world, Dr. Lamorak; you judge us by two-dimensional standards. The surface area of Elsevere is only three-quarters that of the state of New York, but that's irrelevant. Remember we can occupy, if we wish, the entire interior of Elsevere. A sphere of fifty miles' radius has a volume of well over half a million cubic miles. If all of Elsevere were occupied by levels fifty feet apart, the total surface area within the planetoid would be 56,000,000 square miles, and that is equal to the total land area of earth. And none of these square miles, doctor, would be unproductive.'

Well, that's for an asteroid 50 miles in radius and, consequently, 100 miles in diameter. An asteroid that is 35 miles in diameter would have only about $1/27$ the volume, and its levels would offer a surface area of only 2,000,000 square miles, which is nevertheless over half the total area of the United States (0·55 USAs, to be exact).

One small 35-mile-diameter asteroid would then offer as much living space as the moderately large Saturnian satellite Iapetus, if, in the latter case, only surface area were considered.

The material hollowed out of an asteroid would not be waste, either. It could be utilised as a source of metal, and of silicates. The only important elements missing would be hydrogen, carbon and nitrogen, and these could be picked up (remember we're viewing the future through rose-coloured glasses) in virtually limitless quantities from the atmosphere of the gas giants, particularly Jupiter.

If we imagine 100,000 asteroids, all more or less hollowed out, we could end with a living space of 200,000,000,000 square miles or 55,000 USAs. This would be more than 150 times as much area as was available on all the surfaces of the solar system (excluding the gas giants, but even including the asteroids).

Suppose the levels within an asteroid could be as densely populated as the United States today. We might then average 100,000,000 as the population of an asteroid, and the total population of all the asteroids would come to 10,000,000,000,000 (ten trillion).

The question is whether such a population can be supported. One can visualise each asteroid a self-sufficient unit, with all matter vigorously and efficiently cycled. (This, indeed, was the background of the story from which I quoted earlier.)

The bottleneck is bound to be the energy supply, since energy is the one thing consumed despite the efficiency with which all else is cycled.

At the present moment, virtually all our energy supply is derived from the sun. (Exceptions are nuclear energy, of course, and energy drawn from tides or hot springs.) The utilisation of solar energy, almost entirely by way of the green plant, is not efficient, since the green plant makes use of only 2 per cent or so of all the solar energy that falls upon the earth. The unutilised 98 per cent is not the major loss, however.

Solar radiation streams out in all directions from the sun, and when it reaches the earth's orbit, it has spread out over a sphere 93,000,000 miles in radius. The surface area of such a sphere is 110,000,000,000,000,000 (a hundred and ten quadrillion) square miles, while the cross-sectional area presented by the earth is only 50,000,000 square miles.

The fraction of solar radiation stopped by the earth is therefore 50,000,000/110,000,000,000,000,000, or just about 1/2,000,000,000 (one two-billionth).

If all the solar radiation could be trapped and utilised with no greater efficiency than it is now on earth, then the population supportable (assuming energy to be the bottleneck) would mount to two billion times the population of the earth or about 6,000,000,000,000,000,000 (six quintillion).

To be sure, the energy requirement per individual is bound to increase, but then efficiency of utilisation of solar energy may increase also and, for that matter, energy can be rationed. Let's keep the six quintillion figure as a talking point.

To utilise all of solar radiation, power stations would be set up in space in staggered orbits at all inclinations to the ecliptic. As more and more energy was required, the station would present larger surfaces, or there would be more of them, until eventually the entire sun would be encased. Every bit of the radiation would strike one or another of the stations before it had a chance to escape from the solar system.

This would create an interesting effect to any intelligent being studying the sun from another star. The sun's visible light would, over a very short period, astronomically speaking, blank out. Radiation wouldn't cease altogether, but it would be degraded. The sun would begin to radiate only in the infrared.

Perhaps this always happens when an intelligent race becomes intelligent enough, and we ought to keep half an eye peeled out for any star that disappears without going through the supernova stage — any that just blanks out. Who knows?

An even more grisly thought can be expounded. From an energy consideration, I said that a human population of six quintillion might be possible.

On the other hand, the total population of the asteroids, at an American population density, was calculated at a mere ten trillion. Population could still increase 600,000-fold, but where would they find the room?

An increase in the density of the population might seem undesirable and, instead, the men of the asteroids might cast envious eyes on other worlds. Suppose they considered a satellite like Saturn's Phoebe, with its estimated diameter of 200 miles. It could be broken up into about two hundred small asteroids with a diameter of 35 miles each. Instead of one satellite with a surface area of 120,000 square miles, there would be numerous asteroids with a total internal area of 400,000,000 square miles.

The gain might not be great with Phoebe, for considerable hollowing out might be carried on upon that satellite even while it was intact. Still, what about the moon, where hollowing would have to be confined to the outermost skin?

It has a greater mass than all the asteroids put together, and if it were broken up, it would form 200,000 asteroids of 35-mile diameter. At a stroke, the seating capacity, so to speak, of the human race would be tripled.

One can envisage a future in which, one by one, the worlds of the solar system will be broken into fragments for the use of mankind.

But, of course, earth would be in a special class. It would be the original home of the human race, and sentiment might keep it intact.

Once all the bodies of the solar system, except for the gas giants and earth, are broken up, the total number of asteroids would be increased roughly ten-million-fold, and the total human population can then reach the maximum that the energy supply will allow.

But, and here is the crucial point, Pluto may offer difficulties. For one thing, we aren't too certain of its nature. Perhaps its makeup is such that it isn't suitable for breaking up into asteroids. Then, too, it is quite distant. Is it possible that it is too far away for energy to be transmitted efficiently from the solar stations to all the millions of asteroids that can be created from Pluto, out four billion miles from the sun?

If Pluto is ignored, then there is only one way in which mankind can reach its full potential, and that would be to use the earth.

I can see a long drawn-out campaign between the Traditionalists and the Progressives. The former would demand that the earth be kept as a museum of the past and would point out that it was not important to reach full potential population, that a few trillion more or less people didn't matter.

The Progressives would insist that the earth was made for man and not vice versa, that mankind had a right to proliferate to the maximum, and that in any case, the earth was in complete darkness because the solar stations between itself and the sun soaked up virtually all radiation, so that it could scarcely serve as a realistic museum of the past.

I have a feeling that the Progressives would, in the end, win, and I pull down the curtain as the advancing work-fleet, complete with force beams, prepares to make the preliminary incision that will allow the earth's internal heat to blow it apart as the first step in asteroid formation.

CHAPTER SEVEN

ROUND AND ROUND AND...

Anyone who writes a book on astronomy for the general public eventually comes up against the problem of trying to explain that the Moon always presents one face to the earth, but is nevertheless rotating.

To the average reader who has not come up against this problem before and who is impatient with involved subtleties, this is a clear contradiction in terms. It is easy to accept the fact that the Moon always presents one face to the Earth because even to the naked eye, the shadowy blotches on the Moon's surface are always found in the same position. But in that case it seems clear that the Moon is not rotating, for if it were rotating we would, bit by bit, see every portion of its surface.

Now it is no use smiling gently at the lack of sophistication of the average reader, because he happens to be right. The Moon is *not* rotating with respect to the observer on the Earth's surface. When the astronomer says that the Moon *is* rotating, he means with respect to other observers altogether.

For instance, if one watches the Moon over a period of time, one can see that the line marking off the sunlight from the shadow progresses steadily around the Moon; the Sun shines on every portion of the Moon in steady progression. This means that to an observer on the surface of the Sun (and there are very few of those), the Moon would seem to be rotating, for the observer would, little by little, see every portion of the Moon's surface as it turned to be exposed to the sunlight.

But our average reader may reason to himself as follows: 'I see only one face of the Moon and I say it is *not* rotating. An observer on the Sun sees all parts of the Moon and he says it *is* rotating. Clearly, I am more important than the Sun observer since, firstly, I exist and he doesn't, and, secondly, even if he existed, I am me and he isn't. Therefore, I insist on having it my way. The Moon does *not* rotate!'

There has to be a way out of this confusion, so let's think things through a little more systematically. And to do so, let's start with the rotation of the Earth itself, since that is a point nearer to all our hearts.

One thing we can admit to begin with: To an observer on the Earth, the Earth is not rotating. If you stay in one place from now till doomsday, you will see but one portion of the Earth's surface and no other. As far as you are concerned, the planet is standing still. Indeed, through most of civilised human history, even the wisest of men insisted that 'reality' (whatever that may be) exactly matched the appearance and that the Earth 'really' did not rotate. As late as 1633, Galileo found himself in a spot of trouble for maintaining otherwise.

But suppose we had an observer on a star situated (for simplicity's sake) in the plane of the Earth's equator; or, to put it another way, on the celestial equator. Let us further suppose that the observer was equipped with a device that made it possible for him to study the Earth's surface in detail. To him, it would seem that the Earth rotated, for little by little he would see every part of its surface pass before his eyes. By taking note of some particular small feature (for example, you and I standing on some point on the equator) and timing its return, he could even determine the exact period of the Earth's rotation — that is, as far as he is concerned.

We can duplicate his feat, for when the observer on the star sees us exactly in the centre of that part of Earth's surface visible to himself, we in turn see the observer's star directly overhead. And just as he would time the periodic return of ourselves to that centrally located position, so we could time the return of his star to the overhead point. The period determined will be the same in either case. (Let's measure this time in minutes, by the way. A minute can be defined as 60 seconds, where 1 second is equal to 1/31,556,925·9747 of the tropical year.)

The period of Earth's rotation with respect to the star is just about 1,436 minutes. It doesn't matter which star we use, for the apparent motion of the stars with respect to one another, as viewed from the Earth, is so vanishingly small that the constellations can be considered as moving all in one piece.

The period of 1,436 minutes is called Earth's 'sidereal day'. The word 'sidereal' comes from a Latin word for 'star', and the phrase therefore means, roughly speaking, 'the star-based day'.

Suppose, though, that we were considering an observer on the Sun. If he were watching the Earth, he, too, would observe it rotating, but the period of rotation would not seem the same to him as to the observer on the star. Our solar observer would be much closer to the Earth; close enough, in fact, for Earth's motion about the Sun to introduce a new factor. In the course of a single rotation of the Earth (judging by the star's observer), the Earth would have moved an appreciable distance through space, and the solar observer would find himself viewing the planet from a different angle. The Earth would have to turn for four more minutes before it adjusted itself to the new angle of view.

We could interpret these results from the point of view of an observer on the Earth. To duplicate the measurements of the solar observer, we on Earth would have to measure the period of time from one passage of the Sun overhead to the next (from noon to noon, in other words). Because of the revolution of the Earth about the Sun, the Sun seems to move from west to east against the background of the stars. After the passage of one sidereal day, a particular star would have returned to the overhead position, but the Sun would have drifted eastward to a point where four more minutes would be required to make it pass overhead. The solar day is therefore 1,440 minutes long, 4 minutes longer than the sidereal day.

Next, suppose we have an observer on the Moon. He is even closer to the Earth and the apparent motion of the Earth against the stars is some thirteen times greater for him than for an observer on the Sun. Therefore, the discrepancy between what he sees and what the star observer sees is about thirteen times greater than is the Sun/star discrepancy.

If we consider this same situation from the Earth, we would be measuring the time between successive passages of the Moon exactly overhead. The Moon drifts eastward against the starry background at thirteen times the rate the Sun does. After one sidereal day is completed, we have to wait a total of 54 additional minutes for the Moon to pass overhead again. The Earth's 'lunar day' is therefore 1,490 minutes long.

We could also figure out the periods of Earth's rotation with respect to an observer on Venus, on Jupiter, on Halley's Comet, on an artificial satellite, and so on, but I shall have mercy and refrain. We can instead summarise the little we do have in Table 17.

By now it may seem reasonable to ask: But which is *the* day? The *real* day?

The answer to that question is that the question is not a reasonable one at all, but quite unreasonable; and that there is no real day, no real period of rotation. There are only different *apparent* periods, the lengths of which depend upon the position of the observer. To use a prettier-sounding phrase, the length of the period of the earth's rotation depends on the frame of reference, and all frames of reference are equally valid.

Table 17

sidereal day	1,436 minutes
solar day	1,440 minutes
lunar day	1,490 minutes

But if all frames of reference are equally valid, are we left nowhere?

Not at all! Frames of reference may be equally valid, but they are usually not equally useful. In one respect, one particular frame of reference may be most useful; in another respect, another frame of reference may be most useful. We are free to pick and choose, using now one, now another, exactly as suits our dear little hearts.

For instance, I said that the solar day is 1,440 minutes long but actually that's a lie. Because the Earth's axis is tipped to the plane of its orbit and because the Earth is sometimes closer to the Sun and sometimes farther (so that it moves now faster, now slower in its orbit), the solar day is sometimes a little longer than 1,440 minutes and sometimes a little shorter. If you mark off 'noons' that are exactly 1,440 minutes apart all through the year, there will be times during the year when the Sun will pass overhead fully 16 minutes ahead of schedule, and other times when it will pass overhead fully 16 minutes behind schedule. Fortunately, the errors cancel out and by the end of the year all is even again.

For that reason it is not the solar day itself that is 1,440 minutes long, but the average length of all the solar days of the year; this average is the 'mean solar day'. And at noon of all but four days a year, it is not the real Sun that crosses the overhead point but a fictitious body called the 'mean Sun'. The mean Sun is located where the real Sun would be if the real Sun moved perfectly evenly.

The lunar day is even more uneven than the solar day, but the sidereal day is really steady affair. A particular star passes overhead every 1,436 minutes virtually on the dot.

If we're going to measure time, then, it seems obvious that the sidereal day is the most useful, since it is the most constant. Where the sidereal day is used as the basis for checking the clocks of the world by the passage of a star across the hairline of a telescope eyepiece, then the Earth itself, as it rotates with respect to the stars, is serving as the reference clock. The second can then be defined as $1/1436 \cdot 09$ of a sidereal day. (Actually, the length of the year is even more constant than that of the sidereal day, which is why the second is now officially defined as a fraction of the tropical year.)

The solar day, uneven as it is, carries one important advantage. It is based on the position of the Sun, and the position of the Sun determines whether a particular portion of the Earth is in light or in shadow. In short, the solar day is equal to one period of light (daytime) plus one period of darkness (night). The average man throughout history has managed to remain unmoved by the position of the stars, and couldn't have cared less when one of them moved overhead; but he certainly couldn't help noticing, and even being deeply concerned, by the fact that it might be day or night at a particular moment; sunrise or sunset; noon or twilight.

It is the solar day, therefore, which is by far the most useful and important day of all. It was the original basis of time measurement and it is divided into exactly 24 hours, each of which is divided into 60 minutes (and 24 times 60 is 1,440, the number of minutes in a solar day). On this basis, the sidereal day is 23 hours 56 minutes long and the lunar day is 24 hours 50 minutes long.

So useful is the solar day, in fact, that mankind has become accustomed to thinking of it as *the* day, and of thinking that the Earth 'really' rotates in exactly 24 hours, where actually this is only its apparent rotation with respect to the Sun, no more 'real' or 'unreal' than its apparent rotation with respect to any other body. It is no more 'real' or 'unreal', in fact, than the apparent rotation of the Earth with respect to an observer on the Earth — that is, to the apparent lack of rotation altogether.

The lunar day has its uses, too. If we adjusted our watches to lose 2 minutes 5 seconds every hour, it would then be running on a lunar day basis. In that case, we would find that high tide (or low tide) came exactly twice a day and at the same times every day — indeed, at twelve-hour intervals (with minor variations).

And extremely useful is the frame of reference of the Earth itself; to wit, the assumption that the Earth is not rotating at all. In judging a billiard shot, in throwing a baseball, in planning a trip cross-country, we never take into account any rotation of the Earth. We always assume the Earth is standing still.

Now we can pass on to the Moon. For the viewer from the Earth, as I said earlier, it does not rotate at all so that its 'terrestrial day' is of infinite length. Nevertheless, we can maintain that the Moon rotates if we shift our frame of

reference (usually without warning or explanation so that
the reader has trouble following). As a matter of fact, we can
shift our plane of reference to either the Sun or the stars so
that not only can the Moon be considered to rotate but to do
so in either of two periods.

With respect to the stars, the period of the Moon's rotation is 27 days, 7 hours, 43 minutes, 11·5 seconds, or 27·3217
days (where the day referred to is the 24-hour mean solar
day). This is the Moon's sidereal day. It is also the period
(with respect to the stars) of its revolution about the Earth,
so it is almost invariably called the 'sidereal month'.

In one sidereal month, the Moon moves about $1/13$ of the
length of its orbit about the Sun, and to an observer on the
Sun the change in angle of viewpoint is considerable. The
Moon must rotate for over two more days to make up for it.
The period of rotation of the Moon with respect to the Sun
is the same as its period of revolution about the Earth with
respect to the Sun, and this may be called the Moon's solar
day or, better still, the solar month. (As a matter of fact, as
I shall shortly point out, it is called neither.) The solar month
is 29 days, 12 hours, 44 minutes, 2·8 seconds long, or 29·5306
days long.

Of these two months, the solar month is far more useful
to mankind because the phases of the Moon depend on the
relative positions of Moon and Sun. It is therefore 29·5306
days, or one solar month, from new Moon to new Moon, or
from full Moon to full Moon. In ancient times, when the
phases of the Moon were used to mark off the seasons, the
solar month became the most important unit of time.

Indeed, great pains were taken to detect the exact day on
which successive new Moons appeared in order that the
calendar be accurately kept. It was the place of the priestly
caste to take care of this, and the very word 'calendar', for
instance, comes from the Latin word meaning 'to proclaim',

because the beginning of each month was proclaimed with much ceremony. An assembly of priestly officials, such as those that, in ancient times, might have proclaimed the beginning of each month, is called a 'synod'. Consequently, what I have been calling the solar month (the logical name) is, actually, called the 'synodic month'.

The farther a planet is from the Sun and the faster it turns with respect to the stars, the smaller the discrepancy between its sidereal day and solar day. For the planets beyond Earth, the discrepancy can be ignored.

For the two planets closer to the Sun than the Earth the discrepancy is very great. Both Mercury and Venus turn one face eternally to the Sun and have no solar day. They turn with respect to the stars, however, and have a sidereal day which turns out to be as long as the period of their revolution about the Sun (again with respect to the stars).[1]

If the various satellites of the Solar System keep one face to their primaries at all times, as is very likely true, their sidereal day would be equal to their period of revolution about their primary.

If this is so I can prepare Table 18 (not quite like any I have ever seen) listing the sidereal period of rotation for each of the 32 major bodies of the Solar System: the Sun, the Earth, the eight other planets (even Pluto, which has a rotation figure, albeit an uncertain one), the Moon, and 21 other satellites. For the sake of direct comparison I'll give the period in minutes and list them in the order of length. After each satellite I shall put the name of the primary in parentheses and give a number to represent the position of that satellite, counting outwards from the primary.

These figures represent the time it takes for stars to make a complete circuit of the skies from the frame of reference of an observer on the surface of the body in question. If you divide each figure by 720, you get the number of minutes it

Table 18

BODY	SIDEREAL DAY (MINUTES)
Venus	350,000^2
Iapetus (Saturn-8)	104,000
Mercury	82,000^2
Moon (Earth-1)	39,300
Sun	35,060
Hyperion (Saturn-7)	30,600
Callisto (Jupiter-5)	24,000
Titan (Saturn-6)	23,000
Oberon (Uranus-5)	19,400
Titania (Uranus-4)	12,550
Ganymede (Jupiter-4)	10,300
Pluto	8,650
Triton (Neptune-1)	8,450
Rhea (Saturn-5)	6,500
Umbriel (Uranus-3)	5,950
Europa (Jupiter-3)	5,100
Dione (Saturn-4)	3,950
Ariel (Uranus-2)	3,630
Tethys (Saturn-3)	2,720
Io (Jupiter-2)	2,550
Miranda (Uranus-1)	2,030
Enceladus (Saturn-2)	1,975
Deimos (Mars-2)	1,815
Mars	1,477
Earth	1,436
Mimas (Saturn-1)	1,350
Neptune	948
Amaltheia (Jupiter-1)	720
Uranus	645
Saturn	614
Jupiter	590
Phobos	460

would take a star (in the region of the body's celestial equator) to travel the width of the Sun or Moon as seen from the Earth.

On Earth itself, this takes about 2 minutes and no more, believe it or not. On Phobos (Mars's inner satellite), it takes only a little over half a minute. The stars will be whirling by at four times their customary rate, while a bloated Mars hangs motionless in the sky. What a sight that would be to see.

On the Moon, on the other hand, it would take 55 minutes for a star to cover the apparent width of the Sun. Heavenly bodies could be studied over continuous sustained intervals nearly thirty times as long as is possible on the Earth. I have never seen this mentioned as an advantage for a Moon-based telescope, but, combined with the absence of clouds or other atmospheric interference, it makes a lunar observatory something for which astronomers ought to be willing to undergo rocket trips.

On Venus, it would take 485 minutes or 8 hours for a star to travel the apparent width of the Sun as we see it. What a fix astronomers could get on the heavens there — if only there were no clouds.

CHAPTER EIGHT

BEYOND PLUTO

In the last two centuries, the Solar System was drastically enlarged three times; once when Uranus was discovered in 1781; then when Neptune was discovered in 1846; finally when Pluto was discovered in 1930.

Are we all through? Is there no more distant planet to be discovered even yet? We can't know for sure, but at least we can speculate. That much is our fundamental human right.

So — What might a possible Tenth Planet be like? To begin with, how far ought it to be from the Sun? For the answer, we'll go back to the eighteenth century.

Back in 1766, a German astronomer, Johann Daniel Titius, devised a scheme to express simply the distances of the planets from the Sun. He did this by starting with a series of numbers, of which the first was 0, the next 3 and each one following double the one before, thus:

$$0, 3, 6, 12, 24, 48, 96, 192, 384, 768 \ldots$$

Then he added 4 to each term in the series to get the following:

$$4, 7, 10, 16, 28, 52, 100, 196, 388, 772 \ldots$$

Now represent Earth's mean distance from the Sun as 10 and calculate the mean distance of every other planet in proportion. What happens? Well, we can prepare Table 19 listing Titius's series of numbers and comparing them with the relative mean distances from the Sun of the six planets known in Titius's time. Here's how it would look:

Table 19

TITIUS'S SERIES	RELATIVE DISTANCE	PLANET
4	3·9	(1) Mercury
7	7·2	(2) Venus
10	10·0	(3) Earth
16	15·2	(4) Mars
28		
52	52·0	(5) Jupiter
100	95·4	(6) Saturn

When Titius first announced this, no one paid attention, particularly, except for another German astronomer named Johann Bode. Bode wrote about it in 1772, banging the drums hard on its behalf. Since Bode was much more famous than Titius, this relationship of planetary distances has ever since been referred to as Bode's law, while Titius remains in profound obscurity. (This shows you can't always trust to posterity for appreciation either — a thought which should help sadden us further in our moments of depression.)

Even with Bode pushing, the series of numbers was greeted as nothing more than a bit of numerology, worth an absent smile and a that-was-fun-what-shall-we-play-next? But then, in 1781, an amazing thing happened.

A German-born English astronomer, named Friedrich Wilhelm Herschel (he dropped the Friedrich and changed the Wilhelm to William after becoming an Englishman) was engaged that year in a routine sweeping of the skies with one of the telescopes he had built for himself. On March 13th, 1781, he came across a peculiar star that seemed to show a visible disc, which actual stars do not do under the greatest magnification available at that time (or now either, for that matter). He returned to it night after night and by March 19th, he was certain that it was moving with respect to the stars.

Well, anything with a visible disc and movement against the stars could not be a star, so it had to be a comet. Herschel announced the body as a comet to the Royal Society. But then, as he continued his observations, he couldn't help noting that it wasn't fuzzy like a comet, but had a sharply ending disc like a planet. Moreover, after he had observed it for a few months, he could calculate its orbit, and that turned out to be not strongly elliptical like the orbit of a comet but nearly circular like the orbit of a planet. *And* the orbit lay far outside the orbit of Saturn.

So Herschel announced that he had discovered a new planet. What a sensation! Since the telescope had been invented nearly two centuries before, a number of new objects had been discovered; many new stars and several new satellites for both Jupiter and Saturn; but never, never, never in recorded history had a new planet been discovered.

At one bound, Herschel became the most famous astronomer in the world. Within a year he was appointed private astronomer to George III, and six years after that he married a wealthy widow. There was even a move, ultimately defeated, to name the planet he had discovered 'Herschel'. (It is now called Uranus.)

And yet the discovery was accidental and hadn't even been really new. Uranus is actually visible to the naked eye as a very dim 'star', so it was casually seen any number of times. Astronomers had seen it through telescopes and on a number of occasions its position was even reported. As far back as 1690 John Flamsteed, the first British Astronomer Royal, prepared a star map in which he carefully included Uranus — as a star.

In short, any astronomer could have discovered Uranus if he had looked for it. And he would have had a good hint as to what kind of a body to look for and how fast he might expect it to move against the stars, for he would have known its distance from the Sun in advance. Bode's law would have told him. The Bode's law figure for the relative distance of the Seventh Planet (on an Earth-equals-10·0 scale) is 196 and Uranus's actual distance is 191·8.

Obviously, astronomers weren't going to make this mistake again. Bode's law was suddenly the guide to fame and new knowledge and they were going to give it all they had. To begin with, there was that missing planet between Mars and Jupiter. At least *now* they realised there must be a missing planet, for Bode's law had number 28 between the orbits of Mars and Jupiter and no planet was known to exist there. It had to be searched for.

In 1800, twenty-four German astronomers set up a kind of community effort to find the planet. They divided the sky into twenty-four zones and each member was assigned one zone. But alas for planning, efficiency, and Teutonic thoroughness. While they were making all possible preparations, an Italian astronomer, Giuseppe Piazzi, in Palermo, Sicily, accidentally discovered the planet.

It was named Ceres, after the tutelary goddess of Sicily, and proved to be a small object only 485 miles in diameter. It turned out to be only the first of many hundreds of tiny planets ('planetoids') discovered in the region between Mars and Jupiter in the years since. Planetoids numbers 2, 3, and 4, by the way, were found by the German team of astronomers within a year or two after Piazzi's initial discovery, so teamwork wasn't a dead loss after all. Ceres is far the largest of all the planetoids, however, so let's concentrate on it. Its relative mean distance from the Sun is 27·7; Bode's law, as I said, calls for 28.

No astronomer was in the mood to question Bode's law after that.

In fact, when Uranus's motion in its orbit seemed to be a bit irregular, a couple of astronomers, John Couch Adams of England and Urbain J. J. Leverrier of France, independently decided there must be a planet beyond Uranus with a gravitational pull on Uranus that wasn't being allowed for. In 1845 and 1846 they both calculated where the theoretical Eighth Planet ought to be to account for the deviations in Uranus's motions. They did that by beginning with the assumption that its distance from the Sun would be that which was predicted by Bode's law. A few more assumptions and both pointed to the same general position of the sky. And the Eighth Planet, Neptune, proved to be there, indeed.

The only trouble was that it turned out they had made the wrong basic assumption. Neptune ought to have been at relative distance 388 from the Sun. It wasn't; it was at relative distance 301. It was a little matter of 800,000,000 miles closer to the Sun than it should have been, and with one blow that killed Bode's law deader than a dried herring. It went back to being nothing more than an interesting piece of numerology.

When, in 1931, the Ninth Planet, Pluto, was discovered, no one expected it to be at the Bode's-law distance predicted for the Ninth Planet (the numbers of the planets are selected, by the way, by skipping the planetoids so that Mars is the Fourth and Jupiter the Fifth), and it wasn't.

But now wait.

There are four known bodies lying beyond Uranus and every one of them is odd, in one way or another. The four are Neptune and Pluto, plus Neptune's two known satellites, Triton and Nereid.

The oddness of Neptune is, of course, that it lies so much closer to the Sun than Bode's law would indicate. The oddness of Pluto is more complicated. In the first place it has the most eccentric orbit of any of the major planets. At aphelion it recedes to a distance of 4,567,000,000 miles from the Sun, while at perihelion it approaches to a distance of a mere 2,766,000,000 miles. At perihelion it is actually an average of about 25,000,000 miles closer the Sun than is Neptune.

Right now, Pluto is approaching perihelion, which it will reach in 1989. For a couple of decades at the end of the twentieth century, Pluto will remain closer to the Sun than Neptune, then it will move out beyond Neptune's orbit, heading towards its aphelion, which it will reach in 2113.

A second odd feature about Pluto is that the plane of its orbit is tilted sharply to the ecliptic (which is the plane of the Earth's orbit). The tilt is 17 degrees, which is much higher than that of any other planet. It is this tilt which keeps Pluto from ever colliding with Neptune. Although their orbits seem to cross in the usual two-dimensional representation of the Solar System, Pluto is many millions of miles higher than Neptune at the point of apparent crossing.

Finally, Pluto is peculiar in its size. It is 3,600 miles in diameter, much smaller than the four other outer planets. It is also much denser. In fact, in size and mass, it resembles an inner planet such as Mars or Mercury much more than it does any of the outer planets.

Now let's consider Neptune's satellites. One of them, Nereid, is a small thing, 200 miles in diameter and not discovered until 1949. The odd thing about it is the eccentricity of its orbit. At its nearest approach to Neptune, it comes to within 800,000 miles of the planet, then it goes swooping outward to an eventual distance of 6,000,000 miles at the other end of its orbit. Nereid's orbit is by far the most eccentric orbit of any satellite in the Solar System. No planet or planetoid can compare with it either in that respect; only comets equal or exceed that eccentricity.

In contrast to Nereid, Triton is a large satellite, with a diameter in excess of 3,000 miles (as compared with the 2,160 mile diameter of the Moon) and with a nearly circular orbit. The odd thing about it though is that its orbit is tilted sharply to the plane of Neptune's equator; it is quite near to being perpendicular to that plane, in fact.

Now, there are other satellites in the system with eccentric orbits and tilted orbits. They include the seven outermost satellites (unnamed) of Jupiter; and Phoebe, the ninth and outermost satellite of Saturn. Astronomers agree that these outer satellites of Jupiter and Saturn are probably captured planetoids and not original members of the planetary family. The original members (such as the five inner satellites of Jupiter, including the giant satellites, Ganymede, Io, Callisto, and Europa; the eight inner satellites of Saturn, including the giant satellite Titan) all revolve in nearly circular orbits and in the plane of their planet's equator. So, for that matter, do the five small satellites of Uranus and the two small satellites of Mars. From the manner in which satellite systems are supposed to have originated, these circular, untilted orbits seem inevitable.

Well, perhaps Nereid represents a captured planetoid, although it is surprising that a planetoid is to be found so far beyond the planetoid belt, especially one so large (there are not more than four or five planetoids, at most, that are as large as Nereid). And as for Triton, was it captured too? What would an object as large as Triton be doing wandering around in the region of Neptune, getting captured?

Some astronomers have suggested that a catastrophe took place, during some past age, in the neighbourhood of

Neptune. They suggest that Pluto, which is much more nearly the size of a satellite than the size of an outer planet, was originally indeed a satellite of Neptune. However, it was somehow jarred out of position and took up its present wild and eccentric but independently planetary orbit. The shock of that catastrophe may also have jarred Triton's orbit into a strong tilt.

But what was the catastrophe? That no one says.

The one obvious sign of a possible catastrophe in the Solar System is, of course, the asteroid belt. There is no real evidence that there ever was a single planet there, but certainly it is tempting to believe that one was there once and that it exploded (due to the tidal forces within its crust induced by its next-door neighbour, the giant planet, Jupiter, perhaps) An explosion which produced thousands of fragments of rock including Ceres, which is 485 miles in diameter, and three or four others of 100 miles in diameter or more would certainly be a catastrophe.

One catch, however, is that the total mass of all the planetoids between Mars and Jupiter cannot possibly be more than a tenth that of Mars, or more than a fifth that of Mercury. It would still have been far and away the smallest planet in the system. Why should that be? Was it because its neighbour Jupiter gobbled up most of the raw materials for planet formation, leaving our mythical planet a pygmy.

Or suppose that just a fraction of the original planet remained in the space between the orbits of Mars and Jupiter after the explosion? What if the '4½th planet' (we must call it this since Mars is the 4th and Jupiter the 5th) sent a large piece of itself flying far out into space. We can imagine such a piece sailing far out beyond Jupiter, Saturn, Uranus; being caught or seriously deflected by Neptune.

Perhaps the piece was caught by Neptune in an odd orbit and became Triton, while Pluto, as Neptune's original satellite, was knocked out into an independent but whimsical planetary orbit as a result. Or perhaps the piece of the 4½th planet was deflected into the planetary orbit, becoming Pluto, while its gravitational pull tilted Triton's orbit. Or perhaps all three, Pluto, Triton, and Nereid are fragments of the 4½th planet.

The chief bother in all this is how an explosion of the 4½th planet could send so much material far outward, all in one direction. Could it be that this was balanced by the sending of a roughly equal mass inward, towards the Sun?

This brings up the question of our own Moon. Like Triton, the Moon is tilted to the plane of its primary's equator; not by as much, but by a good 18° and its orbit is moderately eccentric as well. Furthermore, the Moon is far too large for us. A planet the size of the Earth has no business with such a huge moon. Of the other inner planets, Mars has two peewee satellites of no account whatever, while Venus and Mercury have none at all.

The Moon is $1/_{80}$ the mass of the Earth and no other satellite in the system even approaches a mass that large in comparison to its primary.

Is it possible then that the inward-speeding fragment of the 4½th planet was captured by the Earth and became the Moon? It sounds, I admit, very unlikely — but speculation is free. Suppose the Moon fragment split up further as it approached Earth and underwent the stresses of our planet's gravitational field. One piece of the fragment might have been slowed sufficiently to allow capture by the Earth, while the other moved at a speed that allowed it to escape from the Solar System altogether.

Or perhaps, to pile improbability upon improbability, this last piece did not escape but was captured by the Sun, so to speak, and became Mercury, which has, next to Pluto, the most eccentric and the most tilted orbit of all the major planets.

If the Moon, Triton, Pluto, and Mercury are all lumped together with the debris of planetoids that are left in the original orbit, you would have a body which would be nearly twice as massive as Mars. This is a respectable planet that would fit the 4½th position nicely.

Of course, I can't imagine what in all this would account for the fact that Neptune's orbit is so much closer to the Sun than it ought to be, but what the devil, we can't have everything. Let's just leave explanations of the fine points to the astronomers and continue to content ourselves with the heady delight of ungoverned speculation. We can suppose that all the bodies beyond Uranus form one complex, to be counted as a single planet, in which the average relationship to the Sun remains what it ought to be, but in which the relationship of the individual pieces has been confused by catastrophe.

If we take the mean distance of the whole complex, that turns out to be (thanks to Pluto) 3,666,000,000 miles which, on the Earth-equals-10·0 basis, comes out to be 395.

Now let's make up a new and more complete Titius series, as in Table 20:

Table 20

TITIUS'S SERIES	RELATIVE DISTANCE	PLANET
4	3·9	(1) Mercury
7	7·2	(2) Venus
10	10·0	(3) Earth
16	15·2	(4) Mars
28	27·7	(4½) Ceres
52	52·0	(5) Jupiter
100	95·4	(6) Saturn
196	191·8	(7) Uranus
388	395	(8, 9) Neptune-Pluto
772	?	(10) Tenth Planet

There you are, then. To answer the question I asked at the beginning of the article, the Tenth Planet should be at position 772 which means it would have a mean distance of 7,200,000,000 miles from the Sun.

How big would it be? Well, if we ignore the interloping Pluto and just consider the other four outer planets, we find a steady decrease in diameter as we move out from Jupiter. The diameters are 86,700 (Jupiter), 71,500 (Saturn), 32,000 (Uranus) and 27,600 (Neptune). Carry that through and let's say the Tenth Planet has a diameter of 10,000 miles, which makes a nice round figure.

With that diameter and at that distance from the Sun (and from us) the Tenth Planet ought to have an apparent magnitude of 13, which would make it rather brighter than the nearer but smaller Pluto. It would show very little disc, but what disc there was would be larger than that of the nearer but smaller Pluto. Well, then, since Pluto has been discovered and the presumably larger and brighter Tenth Planet has not, does that mean the Tenth Planet does not exist?[1]

Not necessarily. Pluto was recognised among a veritable flood of stars of its magnitude or brighter by the fact that it moved among them. So would the Tenth Planet, but at a much slower rate. From Kepler's third law, we can calculate that the period of revolution of the Tenth Planet would be 680 years, nearly three times the length of Pluto's period of revolution, and so the Tenth Planet would move at only one third the rate at which Pluto moves against the stars. It would take a full year for the Tenth Planet to shift its position over the width of the full Moon. This is not the kind of motion that is easily observed by a casual survey of the heavens. So perhaps it has already been seen a number of times and not noticed, as Uranus was.

The thing that strikes me as most unusual about the Tenth Planet is its utter isolation. It is twice as far from Neptune, at Neptune's closest approach to it, as we on Earth are. Most of the time, it is further from Pluto than we are. Once every 2,700 years, allowing the most favourable conditions, Pluto would approach within two and a half billion miles of the Tenth Planet (the distance from Earth to Neptune). Nothing else, barring a possible satellite or comet, would ever come within four and a half billion miles of it.

The Sun would have no discernible disc to the naked eye, of course. It would seem completely starlike and no larger in appearance than the planet Mars appears to us at the time of its closest approach. However, although the Sun would be but a point of light, it would still be over sixty times as bright as our full Moon, and a million times brighter than Sirius, the next brightest object in the sky.

If there were any sentient beings native to the Tenth Planet, that alone ought to tell them there was something different about this particular star. Furthermore, if they watched closely, they would see that the Sun constantly, if slowly, shifted position against the other stars.

As to the planets, all the known members of the Solar System, as seen from the Tenth Planet, would seem to hug the Sun. Even Pluto, viewed from so far beyond its own orbit, would never depart more than 40° from the Sun, even when it happened to be at aphelion at the time of maximum elongation. All other planets would remain far closer to the Sun at all times.

As seen from the Tenth Planet, Mercury and Venus would never be more distant from the Sun than the diameter of our full moon. The Earth would recede, at times, to a distance that was at most half again the width of the full moon and Mars would periodically recede to a distance twice the width of the full moon. I feel certain that, even in the absence of an obscuring atmosphere, all four planets would be lost in the brilliance of the point-size Sun and would never be seen from the Tenth Planet without special equipment.

That leaves only the five outer planets, Jupiter, Saturn, Uranus, Neptune, and Pluto. They would be best seen when well to one side of the Sun at which time they would show up (in telescopes) as fat crescents. In that position, Jupiter, Saturn, Uranus, and Neptune would all be at roughly the same distance from the Tenth Planet. Pluto might, under favourable conditions, be rather closer than the rest.

This means that, with the distance factor eliminated, Saturn would be dimmer than Jupiter, since Saturn is smaller and more distant from the Sun, hence less brightly illuminated. By the same reasoning, Uranus would be dimmer than Saturn, Neptune would be dimmer than Uranus, and Pluto dimmer than Neptune.

In fact, Uranus, Neptune, and Pluto, although approaching more closely to the Tenth Planet at inferior conjunction, than do Jupiter and Saturn, would be invisible to the naked eye.

Jupiter and Saturn would be the only planets visible from the Tenth Planet without special equipment and they would be anything but spectacular. At its brightest, Jupiter would have a magnitude of something like 1·5, about that of the star Castor. And it would only be for a year or so, every six years, that it would approach that brightness, and then it would be only 4° from the Sun and probably not too easy to observe. As for Saturn, there would be two-year periods every fifteen years when it might climb to a brightness of 3·5, about that of an average star. That's all.

Undoubtedly, any astronomers stationed on the Tenth Planet would completely ignore the planets. Any other world in the system would give them a better view. But they would watch the stars. The Tenth Planet would offer them the largest parallaxes in the system, because of its mighty orbital sweep. (Of course, they would have to wait 340 years to get the full parallax.) Measurements of stellar distance by parallax, the most reliable of all methods for the purpose, could be extended one hundred times deeper into space than is now possible.

One last point. What ought we to name the Tenth Planet? We've got to stick to classical mythology by long and revered custom. With the Ninth Planet named Pluto, there might be a temptation to name the Tenth after his consort Proserpina, but that temptation must be resisted. Proserpina is the inevitable name for any satellite of Pluto's that may ever be discovered and should be rigidly reserved for that.

However, consider that the Greeks had a ferryman that carried the souls of the dead across into Hades, the abode of Pluto and Proserpina. His name was Charon. There was also a three-headed dog guarding the entrance of Hades, and its name was Cerberus.

My suggestion then is that the Tenth Planet be named Charon and that its first discovered satellite be named Cerberus.

And then any interstellar voyager returning home and approaching the Solar System on the plane of the ecliptic would have to cross the orbit of Charon and Cerberus to reach the orbit of Pluto and Proserpina. What could be more neatly symbolic than that?

CHAPTER NINE

JUST MOONING AROUND

Almost every book on astronomy I have ever seen, large or small, contains a little table of the Solar System. For each planet, there's given its diameter, its distance from the sun, its time of rotation, its albedo, its density, the number of its moons, and so on.

Since I am morbidly fascinated by numbers, I jump on such tables with the perennial hope of finding new items of information. Occasionally, I am rewarded with such things as surface temperature or orbital velocity, but I never really get enough.

So every once in a while, when the ingenuity-circuits in my brain are purring along with reasonable smoothness, I deduce new types of data for myself out of the material on hand, and while away some idle hours. (At least I did this in the long-gone days when I had idle hours.)

I can still do it, however, provided I put the results into formal essay-form; so come join me and we will just moon around together in this fashion, and see what turns up.

Let's begin this way, for instance . . .

According to Newton, every object in the universe attracts every other object in the universe with a force (f) that is proportional to the product of the masses (m_1 and m_2) of the two objects divided by the square of the distance (d)

between them centre to centre. We multiply by the gravitational constant (g) to convert the proportionality to an equality, and we have:

$$f = \frac{gm_1m_2}{d^2} \qquad \text{(Equation 12)}$$

This means, for instance, that there is an attraction between the Earth and the Sun, and also between the Earth and the Moon and between the Earth and each of the various planets and, for that matter, between the Earth and any meteorite or piece of cosmic dust in the heavens.

Fortunately, the Sun is so overwhelmingly massive compared with everything else in the Solar System that in calculating the orbit of the Earth, or of any other planet, an excellent first approximation is attained if only the planet and the Sun are considered, as though they were alone in the Universe. The effect of other bodies can be calculated later for relatively minor refinements.

In the same way, the orbit of a satellite can be worked out first by supposing that it is alone in the Universe with its primary.

It is at this point that something interests me. If the Sun is so much more massive than any planet, shouldn't it exert a considerable attraction on the satellite even though it is at a much greater distance from that satellite than the primary is? If so, just how considerable is 'considerable'?

To put it another way, suppose we picture a tug of war going on for each satellite, with its planet on one side of the gravitational rope and the Sun on the other. In this tug of war, how well is the Sun doing?

I suppose astronomers have calculated such things, but I have never seen the results reported in any astronomy text, or the subject even discussed, so I'll do it for myself.

Here's how we can go about it. Let us call the mass of a satellite m, the mass of its primary (by which, by the way, I mean the planet it circles) m_p, and the mass of the Sun m_s. The distance from the satellite to its primary will be d_p and the distance from the satellite to the Sun will be d_s. The gravitational force between the satellite and its primary would be f_p and that between the satellite and the Sun would be f_s — and that's the whole business. I promise to use no other symbols in this chapter.

From Equation 12, we can say that the force of attraction between a satellite and its primary would be:

$$f_p = \frac{gmm_p}{d_p{}^2} \qquad \text{(Equation 13)}$$

while that between the same satellite and the Sun would be:

$$f_s = \frac{gmm_s}{d_s{}^2} \qquad \text{(Equation 14)}$$

What we are interested in is how the gravitational force between satellite and primary compares with that between satellite and Sun. In other words we want the ratio f_p/f_s, which we can call the 'tug-of-war value'. To get that we must divide equation 13 by equation 14. The result of such a division would be:

$$f_p/f_s = (m_p/m_s) \ (d_s/d_p)^2 \qquad \text{(Equation 15)}$$

In making the division, a number of simplifications have taken place. For one thing the gravitational constant has dropped out, which means we won't have to bother with an inconveniently small number and some inconvenient units. For another, the mass of the satellite has dropped out. (In other words, in obtaining the tug-of-war value, it doesn't matter how big or little a particular satellite is. The result would be the same in any case.)

What we need for the tug-of-war value (f_p/f_s), is the ratio of the mass of the planet to that of the sun (m_p/m_s) and the square of the ratio of the distance from satellite to Sun to the distance from satellite to primary $(d_s/d_p)^2$.

There are only six planets that have satellites and these, in order of decreasing distance from the Sun, are: Neptune, Uranus, Saturn, Jupiter, Mars, and Earth. (I place Earth at the end, instead of at the beginning, as natural chauvinism would dictate, for my own reasons. You'll find out.)

For these, we will first calculate the mass-ratio and the results turn out as in Table 21.

As you see, the mass ratio is really heavily in favour of the Sun. Even Jupiter, which is by far the most massive planet, is not quite one-thousandth as massive as the Sun. In fact, all the planets together (plus satellites, planetoids, comets, and meteoric matter) make up no more than 1/750 of the mass of the Sun.

Table 21

PLANET	RATIO OF MASS OF PLANET TO MASS OF SUN
Neptune	0·000052
Uranus	0·000044
Saturn	0·00028
Jupiter	0·00095
Mars	0·00000033
Earth	0·0000030

So far, then, the tug of war is all on the side of the Sun. However, we must next get the distance ratio, and that favours the planet heavily, for each satellite is, of course, far closer to its primary than it is to the Sun. And what's more, this favourable (for the planet) ratio must be squared, making it even more favourable, so that in the end we can be reasonably sure that the Sun will lose out in the tug of war. But we'll check, anyway.

Let's take Neptune first. It has two satellites, Triton and Nereid. The average distance of each of these from the Sun is, of necessity, precisely the same as the average distance of Neptune from the Sun, which is 2,797,000,000 miles. The average distance of Triton from Neptune is, however, only 220,000 miles, while the average distance of Nereid from Neptune is 3,460,000 miles.

If we divide the distance from the Sun by the distance from Neptune for each satellite and square the result we get 162,000,000 for Triton and 655,000 for Nereid. We multiply each of these figures by the mass-ratio of Neptune to the Sun, and that gives us the tug-of-war value, in Table 22:

Table 22

SATELLITE	TUG-OF-WAR RATIO
Triton	8,400
Nereid	34

The conditions differ markedly for the two satellites. The gravitational influence of Neptune on its nearer satellite, Triton, is overwhelmingly greater than the influence of the Sun on the same satellite. Triton is firmly in Neptune's grip. The outer satellite, Nereid, however, is attracted by Neptune considerably, but *not* overwhelmingly, more strongly than by the Sun. Furthermore, Nereid has a highly eccentric orbit, the most eccentric of any satellite in the system. It approaches

to within 800,000 miles of Neptune at one end of its orbit
and recedes to as far as 6 million miles at the other end.
When most distant from Neptune, Nereid experiences a
tug-of-war value as low as 11!

For a variety of reasons (the eccentricity of Nereid's orbit,
for one thing) astronomers generally suppose that Nereid is
not a true satellite of Neptune, but a planetoid captured by
Neptune on the occasion of a too-close approach.

Neptune's weak hold on Nereid certainly seems to support
this. In fact, from the long astronomic view, the association
between Neptune and Nereid may be a temporary one.
Perhaps the disturbing effect of the solar pull will eventually
snatch it out of Neptune's grip. Triton, on the other hand,
will never leave Neptune's company short of some catastrophe on a System-wide scale.

There's no point in going through all the details of the
calculations for all the satellites. I'll do the work on my own
and feed you the results. Uranus, for instance, has five
known satellites, all revolving in the plane of Uranus's
equator and all considered true satellites by astronomers.
Reading outward from the planet, they are: Miranda, Ariel,
Umbriel, Titania, and Oberon.

The tug-of-war values for these satellites are given in
Table 23:

Table 23

SATELLITE	TUG-OF-WAR RATIO
Miranda	24,600
Ariel	9,850
Umbriel	4,750
Titania	1,750
Oberon	1,050

All are safely and overwhelmingly in Uranus's grip, and the high tug-of-war values fit with their status as true satellites.

We pass on, then, to Saturn, which has ten satellites: Janus,[1] Mimas, Enceladus, Tethys, Dione, Rhea, Titan, Hyperion, Iapetus, and Phoebe. Of these, the eight innermost revolve in the plane of Saturn's equator and are considered true satellites. Phoebe, the ninth, has a highly inclined orbit and is considered a captured planetoid.

The tug-of-war values for these satellites are given in Table 24:

Table 24

SATELLITE	TUG-OF-WAR RATIO
Janus	23,000
Mimas	15,500
Enceladus	9,800
Tethys	6,400
Dione	4,150
Rhea	2,000
Titan	380
Hyperion	260
Iapetus	45
Phoebe	$3\frac{1}{2}$

Note the low value for Phoebe.

Jupiter has twelve satellites and I'll take them in two instalments. The first five: Amaltheia, Io, Europa, Ganymede, and Callisto, all revolve in the plane of Jupiter's equator and all are considered true satellites. The tug-of-war values for these are given in Table 25:

Table 25

SATELLITE	TUG-OF-WAR RATIO
Amaltheia	18,200
Io	3,260
Europa	1,260
Ganymede	490
Callisto	160

and all are clearly in Jupiter's grip.

Jupiter, however, has seven more satellites which have no official names, and which are commonly known by Roman numerals (from VI to XII) given in the order of their discovery. In order of distance from Jupiter, they are VI, X, VII, XII, XI, VIII, and IX. All are small and with orbits that are eccentric and highly inclined to the plane of Jupiter's equator. Astronomers consider them captured planetoids. (Jupiter is far more massive than the other planets and is nearer the planetoid belt, so it is not surprising that it would capture seven planetoids.)

The tug-of-war results for these seven certainly bear out the captured planetoid notion, as the values given in Table 26 show:

Table 26

SATELLITE	TUG-OF-WAR RATIO
VI	4·4
X	4·3
VII	4·2
XII	1·3
XI	1·2
VIII	1·03
IX	1·03

Jupiter's grip on these outer satellites is feeble indeed.

Mars has two satellites, Phobos and Deimos, each tiny and very close to Mars. They rotate in the plane of Mars's equator, and are considered true satellites. The tug-of-war values are given in Table 27.

So far I have listed 31 satellites, of which 22 are considered true satellites and 9 are usually tabbed as (probably) captured planetoids.[2] I would like, for the moment, to leave out of consideration the 32nd satellite, which happens to be our own Moon (I'll get back to it, I promise) and summarise the 31 in Table 28:

Table 27			Table 28		
	TUG-OF-WAR		NUMBER OF SATELLITES		
SATELLITE	RATIO		PLANET	TRUE	CAPTURED
Phobos	195		Neptune	1	1
Deimos	32		Uranus	5	0
			Saturn	9	1
			Jupiter	5	7
			Mars	2	0

And now let's analyse this list in terms of tug-of-war values. Among the true satellites the lowest tug-of-war value is that of Deimos, 32. On the other hand, among the nine satellites listed as captured, the highest tug-of-war value is that of Nereid with an average of 34.

Let us accept this state of affairs and assume that the tug-of-war figure 30 is a reasonable minimum for a true satellite and that any satellite with a lower figure is, in all likelihood, a captured and probably temporary member of the planet's family.

Knowing the mass of a planet and its distance from the Sun, we can calculate the distance from the planet's centre at which this tug-of-war value will be found. We can use Equation 15 for the purpose, setting f_p/f_s equal to 30, putting in the known values for m_p, m_s, and d_s, and then solving for d_p. That will be the maximum distance at which we can expect to find a true satellite. The only planet that can't be handled in this way is Pluto, for which the value of m_p is very uncertain, but I omit Pluto cheerfully.

We can also set a minimum distance at which we can expect a true satellite; or, at least, a true satellite in the usual form. It has been calculated that if a true satellite is closer to its primary than a certain distance, tidal forces will break it up into fragments. Conversely, if fragments already exist at such a distance, they will not coalesce into a single body. This limit of distance is called the 'Roche limit' and is named for the astronomer E. Roche, who worked it out in 1849. The Roche limit is a distance from a planetary centre equal to 2·44 times the planet's radius.[3]

So, sparing you the actual calculations, here are the results in Table 29 for the four outer planets:

Table 29

DISTANCE OF TRUE SATELLITE
(MILES FROM THE CENTRE OF THE PRIMARY)

PLANET	MAXIMUM (TUG-OF-WAR = 30)	MINIMUM (ROCHE LIMIT)
Neptune	3,700,000	38,000
Uranus	2,200,000	39,000
Saturn	2,700,000	87,000
Jupiter	2,700,000	106,000

As you see, each of these outer planets, with huge masses and far distant from the competing Sun, has ample room for large and complicated satellite systems within these generous limits, and the 22 true satellites all fall within them.

Saturn does possess something within Roche's limit — its ring system. The outermost edge of the ring system stretches out to a distance of 85,000 miles from the planet's centre. Obviously the material in the rings could have been collected into a true satellite if it had not been so near Saturn.

The ring system is unique as far as visible planets are concerned, but of course the only planets we can see are those of our own Solar System. Even of these, the only ones we can reasonably consider in connection with satellites (I'll explain why in a moment) are the four large ones.

Of these, Saturn has a ring system and Jupiter just barely misses one. Its innermost satellite, Amaltheia, is about 110,000 miles from the planet's centre, with the Roche limit at 106,000 miles. A few thousand miles inward and Jupiter would have rings. I would like to make the suggestion therefore that once we reach outward to explore other stellar systems we will discover (probably to our initial amazement) that about half the large planets we find will be equipped with rings after the fashion of Saturn.

Next we can try to do the same thing for the inner planets. Since the inner planets are, one and all, much less massive than the outer ones and much closer to the competing Sun, we might guess that the range of distances open to true satellite formation would be more limited, and we would be right. Here are the actual figures in Table 30 as I have calculated them:

Table 30

DISTANCE OF TRUE SATELLITE
(MILES FROM THE CENTRE OF THE PRIMARY)

PLANET	MAXIMUM (TUG-OF-WAR = 30)	MINIMUM (ROCHE LIMIT)
Mars	15,000	5,150
Earth	29,000	9,600
Venus	19,000	9,200
Mercury	1,300	3,800

Thus, you see, where each of the outer planets has a range of two million miles or more within which true satellites could form, the situation is far more restricted for the inner planets. Mars and Venus have a permissible range of but 10,000 miles. Earth does a little better, with 20,000 miles.

Mercury is the most interesting case. The maximum distance at which it can expect to form a natural satellite against the overwhelming competition of the nearby Sun is well within the Roche limit. It follows from that, if my reasoning is correct, that Mercury *cannot* have a true satellite, and that anything more than a possible spattering of gravel is not to be expected.

In actual truth, no satellite has been located for Mercury but, as far as I know, nobody has endeavoured to present a reason for this or treat it as anything other than an empirical fact. If any Gentle Reader, with a greater knowledge of astronomic detail than myself, will write to tell me that I have been anticipated in this, and by whom, I will try to take the news philosophically. At the very least, I will confine my kicking and screaming to the privacy of my study.

Venus, Earth, and Mars are better off than Mercury and do have a little room for true satellites beyond the Roche limit. It is not much room, however, and the chances of gathering enough material over so small a volume of space to make anything but a very tiny satellite is minute.

And, as it happens, neither Venus nor Earth has any satellite at all (barring possible minute chunks of gravel) within the indicated limits, and Mars has two small satellites, each less than 20 miles across, which scarcely deserve the name.

It is amazing and very gratifying to me, to note how all this makes such delightful sense, and how well I can reason out the details of the satellite systems of the various planets.

It is such a shame that one small thing remains unaccounted for; one trifling thing I have ignored so far, but —

WHAT IN BLAZES IS OUR OWN MOON DOING WAY OUT THERE?

It's too far out to be a true satellite of the Earth, if we go by my beautiful chain of reasoning — which is too beautiful for me to abandon. It's too big to have been captured by the Earth. The chances of such a capture having been effected and the Moon then having taken up a nearly circular orbit about the Earth are too small to make such an eventuality credible.

There are theories, of course, to the effect that the Moon was once much closer to the Earth (within my permitted limits for a true satellite) and then gradually moved away as a result of tidal action. Well, I have an objection to that. If the Moon were a true satellite that originally had circled Earth at a distance of, say, 20,000 miles, it would almost certainly be orbiting in the plane of Earth's equator and it isn't.

But, then, if the Moon is neither a true satellite of the Earth nor a captured one, what is it?[4] This may surprise you, but I have an answer; and to explain what that answer is, let's get back to my tug-of-war determinations. There is, after all, one satellite for which I have not calculated it, and that is our Moon. We'll do that now.

The average distance of the Moon from the Earth is 237,000 miles, and the average distance of the Moon from the Sun is 93,000,000 miles. The ratio of the Moon — Sun distance to the Moon — Earth distance is 392. Squaring that gives us 154,000. The ratio of the mass of the Earth to that of the Sun was given earlier in the chapter and is 0·0000030. Multiplying this figure by 154,000 gives us the tug-of-war value, presented in Table 31:

Table 31

	TUG-OF-WAR
SATELLITE	RATIO
Moon	0·46

The Moon, in other words, is unique among the satellites of the Solar System in that its primary (us) *loses* the tug of war with the Sun. The Sun attracts the Moon twice as strongly as the Earth does.

We might look upon the Moon, then, as neither a true satellite of the Earth nor a captured one, but as a planet in its own right, moving about the Sun in careful step with the Earth. To be sure, from within the Earth-Moon system, the simplest way of picturing the situation is to have the Moon revolve about the Earth; but if you were to draw a picture of the orbits of the Earth and Moon about the Sun exactly to scale, you would see that the Moon's orbit is everywhere concave towards the Sun. It is always 'falling' towards the Sun. All the other satellites, without exception, 'fall' away from the Sun through part of their orbits, caught as they are by the superior pull of their primary — but not the Moon.

And consider this — the Moon does not revolve about the Earth in the plane of Earth's equator, as would be expected of a true satellite. Rather it revolves about the Earth in a plane quite close to that of the ecliptic; that is, to the plane in which the planets, generally, rotate about the Sun. This is just what would be expected of a planet!

Is it possible then, that there is an intermediate point between the situation of a massive planet far distant from the Sun, which develops about a single core, with numerous satellites formed, and that of a small planet near the Sun which develops about a single core with no satellites? Can there be a boundary condition, so to speak, in which there is condensation about two major cores so that a double planet is formed?

Maybe Earth just hit the edge of the permissible mass and distance; a little too small, a little too close. Perhaps if it were better situated the two halves of the double planet would have been more of a size. Perhaps both might have had atmospheres and oceans and — life. Perhaps in other stellar systems with a double planet, a greater equality is more usual.

What a shame if we have missed that . . .

Or, maybe (who knows), what luck!

CHAPTER TEN

STEPPINGSTONES TO THE STARS

There's something essentially unsatisfactory to me about the conquest of the Solar System which now seems to be at hand.[1] We know too much about what we'll find, and what we'll find won't be enough.

After all, except for some possible lichenlike objects on Mars, the other worlds of the Solar System are all barren (barring a most unexpected miracle).

Sure, we'll get all sorts of information and knowledge. In the process of reaching these barren worlds, we'll develop valuable alloys, plastics, fuels. We'll work up useful techniques of miniaturisation, automation, and computation. I wouldn't minimise any of these advances.

But — there will be no Martian princesses, no tentacled menaces, no superhumanly intelligent energy beings, no dreadful monsters to bring back to zoos. In short, there won't be any romance!

For the proper results and rewards of space travel, we must reach the stars. We must find the Earth-type planets that possibly circle them, carrying upon them their full complement (we hope) of friend and foe, of superman and monster.

Only how do we get to the stars? The Moon may be on our doorstep and Mars may be just across the threshold, but the stars are way to helengone out of sight.

The Moon is 222,000 miles away at its nearest and Mars is 35,000,000 miles away at its nearest. Even Pluto, the most distant of the known planets, is never further than 4,650,000,000 miles from us. On the other hand, the Alpha Centauri system, which includes the nearest stars to us, is 25,000,000,000,000 miles away.

In other words, when we've laboured our way to the farthest edge of the Solar System and stand on Pluto, we have covered a distance which is, at best, less than $1/5000$ of the distance that must be covered if even the nearest star is to be reached.

It would be so nice if there were steppingstones to the stars; if there were bodies between Pluto and the stars that would at least give us a breathing spell, a place to stop and rest on the long trip to the nearest stars.

And having said that, I can smile cheerfully and say that there is good reason to believe that such steppingstones do exist. I don't mean dark stars which may or may not exist between us and Alpha Centauri; and I don't mean trans-Plutonian planets, which may or may not exist.

I am referring, rather, to a shell of planetoids which surrounds the Sun, far beyond Pluto's orbit, with a dark halo; a shell of planetoids that dwarfs the known Solar System and which, in all probability, actually exists.

To tell the story of these planetoids, I shall, as is my wont, begin at the beginning. In this case, the beginning involves the comets.

From time immemorial, comets have been considered portents of disaster, and with what seemed good reason.

After all, the heavens are, for the most part, a scene of quiet changelessness or, at most, of majestically periodic change. The sun rises and sets, the moon runs through its phases, the 'fixed' stars maintain their positions exactly from generation to generation, and the planets wander among them in complicated but predictable paths.

All is well. All is peaceful.

Then, hurrying into view, apparently from nowhere, comes a comet. It is like nothing else in the heavens. A fuzzy patch of light, the 'coma', surrounds a bright starlike nucleus, and extending from the coma is an arched tail that can stretch halfway across the heavens. Having come from nowhere, the comet finally vanishes into nowhere. There seemed no way of predicting either its coming or going and all one could say was that it had disturbed the peace and serenity of the skies.

This was in itself disturbing enough. Add to that the strangeness of its shape. It resembled a distraught woman, tearing across the sky in a hysterical frenzy, her unbound hair streaming behind her in the wind. The very word 'comet' comes from the Greek *kometes* meaning 'long-haired'.

Naturally, any sensible man could only suppose that such a sudden and frightening apparition was sent by some god to warn humanity of disaster. And furthermore, since life and humanity are such that disaster strikes every year without fail, this theory seems to be borne out unmistakably. After a comet, disaster invariably follows. Within a year of the comet's appearance, there is sure to be a war, plague, or famine somewhere, or some famous man dies, or some heretic appears or something.

The last halfway spectacular comet showed up in 1910, and it succeeded in frightening many people into believing the end of the world would surely come. (It also, as any fool can plainly see, foretold the death of Mark Twain, the sinking of the *Titanic*, the coming of World War I, and a whole slew of catastrophes.)

However, portent or not, what is the nature of a comet? Aristotle, and the ancient and medieval thinkers who followed him, believed the heavens were perfect and unchangeable. Since comets came and went, having a beginning and an ending (which stars and planets did not) they were imperfect and changeable and, therefore, could not be part of the heavens. They were instead atmospheric phenomena; exhalations of bad air and therefore part of our own corrupt and miserable Earth.

This notion was not destroyed until 1577. The Danish astronomer, Tycho Brahe, measured the parallax of a bright comet that appeared that year, plotting its position as seen against the stars from his own observatory in Denmark and from another observatory in Prague. The parallax proved too small to measure. This is not surprising, considering the relative shortness of the base line (about 500 miles) and the fact that this was before the days of the telescope. However, even so, if the comet had been within 600,000 miles of the Earth, its parallax could have been perceptible. Tycho's conclusion, then, was that the comet had to be *at least* three times as far from the Earth as the Moon was. That made that comet, at any rate, part of the heavens; and Aristotle was wrong.

Even as part of the heavens rather than of the Earth, comets remained troublesome. They didn't fit into any system. When Copernicus put the Sun at the centre of the Solar System and Kepler made planetary orbits into ellipses, the design of the planets began to fall neatly into place — except for the comets. They still came from nowhere, vanished into nowhere, and represented an irritating lawlessness in the kingdom of the Sun.

Then came Newton and his law of gravitation that so neatly explained the planetary movements. Could it also explain cometary movements? That would indeed be an acid test.

In the year 1704 Edmund Halley, a good friend of Newton, began to work out the orbits of various comets over the regions for which observational records existed, in order to see if their motions could be made to fit the requirements of gravitational mathematics. The records of twenty-four different comets were studied.

The one with the best available data was the comet of 1682, which Halley had himself observed. Working out its orbit, he noticed that it passed through the same regions of the sky as had the comet of 1607, seventy-five years before, and the comet of 1531, seventy-six years before that. Checking back, he found records of another comet in 1456, seventy-five years further back still.

Could it be that the same comet was coming back at intervals of seventy-five years or so, after passing over an elliptical orbit so eccentric that its far end reached out way beyond the orbit of Saturn, then the furthest planet known?

Halley felt certain that just this was indeed so, and consequently predicted that the comet of 1682 would return once again in 1758.

It is one of the frustrations of scientific history that Halley knew he was not likely to live to see his prediction verified or exploded. He would have had to live to be one hundred and two for that, and he didn't. He made a valiant try, reaching the age of eighty-five, but that wasn't good enough.

On Christmas night 1758 a comet was sighted and through early 1759, it rode high in the sky. The comet had indeed returned and it has been called Halley's comet ever since. (It was Halley's comet that was in the sky in 1910.)

This created a sensation. Comets, or at least one comet, had been reduced to a commonplace, law-abiding member of the Solar System. Since then, many others have been supplied with definite orbits. And now, at last, there is no logical reason for considering comets divinely sent portents of disaster — which, however, will not prevent people preparing for the end of the world at the next appearance of a large comet, you may be sure.

Granted that comets are ordinary members of the Solar System, subject to the same laws of motion as are the sedate planets, what are they? Well, they aren't much.

Comets have frequently approached one or another of the various planets and have had their orbits altered, sometimes drastically, as a result of the gravitational attraction of the planet. (Such perturbations make it rather difficult to pinpoint the time of a comet's return.) The planet, for its part, has never in any way showed any measurable effect due to the comet's gravitational attraction. The comet of 1779 actually passed through Jupiter's satellite system without affecting the satellites in any way.

The obvious conclusion is that for all their gigantic volumes, and some comets are actually more voluminous than the Sun, comets have very small masses. The mass of even a large comet can be no larger than that of a middle-sized planetoid.

If this is so, the density of a comet must be extremely low, far lower than the density of Earth's atmosphere. This is demonstrated by the fact that stars can be seen through the tail of a comet with no perceptible diminution in brightness. The Earth passed through the tail of Halley's comet in 1910 and there was no discernible effect. In fact, Halley's comet passed between the Earth and the Sun and the whole thing disappeared. The Sun shone through it as though it were a vacuum.

Professor Fred Whipple of Harvard originated, some years ago, a now widely accepted theory of the composition of comets that accounts for all this. Comets, he supposes, are made up largely of 'ices', that is, of low-melting solids such as water, methane, carbon dioxide, ammonia, and so on. When far from the Sun, these substances are indeed solid and the comet is a small, solid body. As it approaches the Sun, however, some of the ices evaporate and the dust and gas that form are forced away from the Sun by the Solar wind.[2]

Sure enough (as was first observed in 1531) a comet's tail always points generally away from the Sun. It streams out behind the comet as the comet approaches the Sun, but it precedes the comet as it moves away from the Sun. Moreover, the closer to the Sun, the larger the tail.

Not as much atmosphere is formed, driven away by radiation pressure and lost, as you might think. The ices themselves are poor conductors of heat and comets remain in the vicinity of the Sun only a comparatively short space of time. They retreat with most of their substance intact.

Nevertheless, at each return a comet does lose some of its substance. Whatever passes into the tail vanishes into space and never returns. A few dozen passes at the Sun would probably suffice to finish a comet. Even a comet that returns only at intervals of a century or so can't be expected to last more than several thousand years at best. Therefore we ought, within historical times, to see comets shrivel and die.

And we do. Halley's comet at its return in 1910 was disappointingly dim, when compared with previous descriptions. It will probably be even more disappointing at its next scheduled appearance in 1986. It is dying.

And some comets have actually died as men watched. The best known example is that of Biela's comet, first discovered in 1772 by the German astronomer, Wilhelm von Biela. It had a period of about 6·6 years and was observed on a number of its returns. In 1846, it was found to have split in two, the halves travelling side by side. In 1852 the two parts had separated further. And Biela's comet was never seen again. It had died.

But that's not the end of the story. Travelling in the orbit of the comet are a group of meteorites. We know, because in 1872 Biela's comet would have passed fairly close to the Earth if there had still been a Biela's comet. There wasn't, but that year we were treated with a meteor shower radiating out of the spot where the comet would have been located.

Apparently, embedded in the ices of the comet are a vast number of pebbles and pinpoints or less of metal and silicates. When the binding ices are gone, the contents fall apart. The small meteors and micrometeors that fill space now may thus be the ghosts of comets long dead.

Obviously, if comets have such short lifetimes and are still as numerous as they are (several new ones are discovered every year) even though the Solar System has been in existence for five billion years, a continual supply must be entering the system. But where are they coming from, then?

The easiest answer is that they come from interstellar space. They may be wanderers among the stars. Some may occasionally enter the gravitational field of the Sun, flash around it and go forever. Some enter, are captured by planets, and become periodic comets, doomed to a quick death.

There are a couple of arguments against this possibility. First, to have interstellar migrants blundering into our Solar System at the rate they do would require the filling of interstellar space with a most unlikely number of comets. Besides, more would enter the system from the direction towards which the Sun is travelling than from the other. That, however, is not so. Comets come from all directions equally.

Secondly, if comets entered the system randomly from outer space, a number should come and go in distinctly hyperbolic orbits (like a hairpin opened wide). No comet with a *distinctly* hyperbolic orbit has ever been observed.

In view of this, a more logical possibility is that the source of the comets is a local reservoir bound to the Sun. It was suggested some years ago that this local reservoir exists in the form of a shell of ice planetoids, located from one to two light-years from the Sun in every direction.

It is easy to see how this shell may have come into existence. If the Solar System began as a vast turbulent cloud of dust and gas some light-years in diameter, then as it swirled and contracted, the planets and present-day Sun would be formed. At the outskirts of the original cloud, however, the density would have been too low for planetary formation and, instead, there would be numerous local concentrations. Since the temperature has remained near the absolute zero throughout billions of years in that far-flung region, the ices which composed much of the original cloud, would be retained even by the tiny gravity of the planetoids. (Nearer the Sun, the higher temperature has caused even as large a body as the Earth to lose much of its supply of ices.)

There is an estimate to the effect that this shell of 'cometary planetoids' contains 100,000,000,000 individuals with a mass, all told, up to $1/100$ or even possibly $1/10$ that of the Earth. The average cometary planetoid would then have a mass of 600,000,000 to 6,000,000,000 tons. If we assumed the density of such a planetoid to be equal to that of ice, the average diameter would run, roughly, close to a mile.

You might think that a shell of a hundred billion planetoids ought somehow to make its presence known to observers on Earth. However, consider that the shell of space enclosing the Sun at a distance between one and two light-years, has a volume of thirty cubic light-years. This is immense! If the 100 billion cometary planetoids were evenly distributed through that volume, the average distance separating them would be about 1¼ billion miles, which is nearly the distance between ourselves and Uranus.

Naturally, a volume of space containing a mile-wide hunk of ice every billion miles or so is not going to make any impression at all at a distance of a light-year or more. The cometary planetoids will reveal themselves neither by luminosity nor by blocking the light of the stars.

Imagine a cometary planetoid somewhere in the middle of the shell, say 1½ light-years from the Sun. The Sun, from that distance, would seem merely a star, though still the brightest star in the sky, with a magnitude of -2. The planetoid would still be within the gravitational influence of the Sun (no other star would be as close) but that influence would be weak.

A cometary planetoid, 1½ light-years from the Sun, and travelling in a circular orbit about the Sun, would be whipped along under the feeble gravitational lash at a speed of only a little over 3 miles a minute. This may sound fast to the automobile driver, but the Earth moves along its orbit at a rate of 1,100 miles a minute and even far-off Pluto never moves at a rate of less than 150 miles a minute.

At its slow rate of movement, it takes the average cometary planetoid 30,000,000 years to complete a revolution about the Sun. In all the existence of the Solar System, those far-distant planetoids have not, on the average, yet had time to revolve about the Sun 200 times.

But if the cometary planetoids are circling quietly way out there, why do they not continue to circle there forever? What sends them down towards the Sun? The only possible answer seems to involve the interfering gravitational influence of the nearer stars. After all, the gravitational pull of Alpha Centauri on those cometary planetoids which happen to be directly between that star and the Sun, is 10 per cent that of the Sun and that is not negligible. (Remember, Alpha Centauri is scarcely farther from some of those planetoids than the Sun is.) A few other stars exert gravitational attractions for those planetoids nearest them to an amount of over 1 per cent that of the Sun.

Now then, if these stellar attractions catch a particular planetoid in such a way as to slow its orbital velocity, it must fall in towards the Sun, its circular orbit becoming elliptical. If the orbital velocity is slowed sufficiently, it must fall in towards the Sun so sharply as to enter the Solar System proper. It would gather speed as it did so, whip around the Sun, and climb back to the point where the perturbation had taken place, then whip down again, climb back, whip down again, and so on. If it came close enough to the Sun, it would develop a gigantic tail and coma of evaporating ices and would become visible to watchers on the Earth.

If only the Sun and the comet existed, this new, highly elliptical orbit would be permanent (barring additional stellar perturbation). A comet travelling in such an orbit would have a much shorter year than it did when it was in its shell, but its year would still be long by Earthly standards — about 10,000,000 years or so.

As far as man is concerned, such 'long-period comets' would be one-shots. Any comet of this type appearing during historical times would not have been viewed by man on its previous visit, for he did not then exist. Moreover there is a distressingly good chance that man may no longer exist to see the next visit.

Of course, once a comet enters the Solar System proper, there is always the chance that it will come close enough to some planet to have its orbit affected. In some cases, its velocity will be speeded so that its orbit will become slightly hyperbolic and it may then leave the Solar System for good. In other cases, its velocity will be slowed and it will no longer gain the kinetic energy required to send it back to the cometary shell. It will often only recede no further than the neighbourhood of the planetary perturbation, so that it will, in effect, have been captured by the planet.

All the outer planets have 'families' of comets, that of Jupiter, very naturally, being the largest. Perhaps the most remarkable of the Jupiter family is Encke's comet, the orbit of which was worked out in 1818 by the German astronomer Johann Franz Encke, after it had been discovered by the French astronomer, Jean Louis Pons.

Encke's comet has the shortest period of any known comet — 3·3 years. It never recedes further from the Sun than about 400,000,000 miles which means that even at its most distant, it is never as far from the Sun as Jupiter is. It approaches fairly close to Mercury's orbit at its perihelion and its perturbation by Mercury has been used to calculate the mass of that small planet.

As you might expect, Encke's comet is dim and unspectacular, and it never develops a tail. It has been near the Sun far too many times to be anything else. Most of its ices are undoubtedly gone and it must now consist largely of a fairly compact silicate residue, thinly interlarded, perhaps, with the remnant of the original ices.

Naturally, the cometary shell is being depleted by these stellar perturbations. Any cometary planetoid slowed and sent down into the Solar System proper is condemned to death. In addition, other cometary planetoids are speeded by stellar perturbations and may be forced into a hyperbolic orbit that drives them away from the Sun altogether.

On the other hand, no cometary planetoids are being added to the shell as far as we know, so that the number continually declines.

However, this need not be a source of worry. It has been estimated that perhaps three new comets are sent hurling into the Solar System proper each year. We can suppose that three more are, on the average, speeded into hyperbolic loss in each year. At that rate, in the entire five-billion-year history of the Solar System, 30,000,000,000 cometary planetoids have been lost or destroyed. That amounts to only 30 per cent of the estimated number that still remains.

Despite the cometary death rate, then, our comets will be with us in their usual numbers for billions of years more.

It is these cometary planetoids, to get back to the remarks I made at the very beginning of the article, which may represent the steppingstones to the stars.

If we could ever reach Pluto, it might not be too great a hop to reach one of the closer cometary planetoids; one that had been slowed into a relatively skimming approach towards the outskirts of the Solar System proper. Certainly not as great an effort would be required to reach such a planetoid as would be required to reach Alpha Centauri in one jump.

If a base could be set up on such a mile-wide hunk of ices, perhaps we could continue to press outward into space from planetoid to planetoid in an island-hopping fashion, to the outermost fringes of the shell.

Nor would the two-light-year mark necessarily end such island-hopping possibilities. After all, there is no reason to believe that Alpha Centauri doesn't have a halo of cometary planetoids of its own. Why shouldn't it have one? (Though perhaps a more complicated one, since Alpha Centauri is really three stars.)

If it has one, then Alpha Centauri and the Sun are close enough so that the outermost fringes of the halo of one ought to be rather close to the outermost fringes of the halo of the other.

Perhaps, then, we could island-hop over the ice all the way. Perhaps at no point will an uninterrupted trip of more than a few billion miles be required and perhaps we can reach the nearest star, at least, in the way a mountain climber scales a high peak — by establishing a series of intermediate bases on the way.

I cannot honestly say that this makes a trip to the stars actually look inviting, but if we've *got* to go, surely it is easiest to go a step at a time.

CHAPTER ELEVEN

THE PLANET OF THE DOUBLE SUN

The title sounds as though this were going to be a rather old-fashioned science-fiction story, doesn't it?

Yet although the title may sound old-fashioned, the situation need not be. One of the most glamorous settings that can be imagined is that of more than one sun in the sky.

The author of a story describing such a setting need not (and usually does not) worry about the astronomic verities of the situation. The suns are usually described as looking like suns and both (or all) are made to move independently in the sky. The author will usually throw in local colour by saying that one sun was just rising, while the other had just passed zenith. He may make matters more colourful (figuratively *and* literally) by having one sun, for example, red and the other blue. Then he can talk of double shadows and their various configurations and colour combinations.[1]

A little of this is enough to make us sigh at our misfortune in having only one sun in the sky; and a pretty colourless one at that. Oh, the missing glories!

What *would* it be like to have more than one sun in the sky? There are, of course, a wide variety of types of multiple stars; some are made up of two components and some of more than two. In some multiple stars, the components are near together; in others far apart. The components may be similar or not similar; one may be a red giant or one may be a white dwarf.

But let's not make up any systems or look for something exotic or foreign. The fact of the matter is that we have an example in our back yard. The nearest star to us in space, a star so close we can almost reach out and touch it, a next-door neighbour no more than 25,000,000,000,000 miles away, good old Alpha Centauri, is a multiple star.

Suppose we were on a planet in the Alpha Centauri system. What would it be like?

To begin with, what is Alpha Centauri like?

Alpha Centauri is a star in the Southern Celestial Hemisphere. It is never visible in the sky north of about 30 degrees north latitude. The chances are you've never seen it; I know I never have. Moreover, the ancient Greeks never saw it.

The chief observatories of the medieval Arabs, in Cordova, Baghdad, and Damascus, were all north of the 30-degree line. Presumably ordinary Arabs in the Arabian and Sahara deserts must occasionally have seen a bright star very near the southern horizon, but this, apparently, did not penetrate to the egghead level.

The test of the matter is that Alpha Centauri, although the third brightest star in the sky, has no name of its own, neither Greek nor Arabic. (The name Alpha Centauri is official 'astronomese'.)

Of course, once Europeans started adventuring down the coast of Africa in the late 1400s, the bright star must have been observed at once. Eventually, astronomers got around to making star maps of those parts of the Southern Celestial Hemisphere invisible from Europe. (The first was Edmund Halley of Halley's comet fame who, in 1676, at the age of twenty, travelled to St. Helena of future Napoleonic fame to map southern stars.) Astronomers divided the southern heavens into constellations to complete the scheme already begun in those parts of the heavens which the ancients had been able to observe.

They named the constellations in Latin, naturally, and included mythological creatures as a further match to what already existed in the sky (just as planets discovered in modern times received mythological names matching the older ones). One of the prominent southern constellations was named the Centaur. In Latin, this is *Centaurus* and the genitive ('of the Centaur') is *Centauri*.

Centaurus contains two first magnitude stars. The brighter was named Alpha Centauri and the other Beta Centauri. The words 'alpha' and 'beta' are not only the first two letters of the Greek alphabet but were also used by the Greeks to represent the numbers 'one' and 'two', a habit never broken by scientists. The names of the stars, freely translated, therefore mean 'star number one of the Centaur' and 'star number two of the Centaur' respectively.

The magnitude of Alpha Centauri is 0·06 which makes it, as said, the third brightest star of the sky. The only stars brighter are Canopus (−0·86) and, of course, Sirius (−1·58).

(The lower the magnitude, the brighter the star, in a logarithmic ratio. A difference in magnitude of one unit means a difference in brightness of 2·512 times. A difference in magnitude of two units means a difference in brightness of 2·512 × 2·512 or about 6·31 times and so on).

About 1650 telescopes became good enough to detect the fact that some stars, which looked like single points of light to the naked eye, were actually two closely spaced points of light. In 1685 Jesuit missionaries in Africa, taking time out for astronomical observations, first noticed that Alpha Centauri is an example of such a double star. The brighter component is Alpha Centauri A, the other Alpha Centauri B.

The magnitude of Alpha Centauri A by itself is 0·3 and that of Alpha Centauri B is 1·7. The 1·4 difference in magnitude means that Alpha Centauri A is 3·6 times as bright as Alpha Centauri B. To translate the brightness into absolute terms; that is, to compare either component with our Sun, it is necessary to know the distance of Alpha Centauri.

This distance could be measured by noting slight shifts in the star's position, mirroring the change in Earth's position as it revolved about the Sun. This tiny yearly motion of a star, resulting from Earth's motion, is called stellar parallax and grows smaller as the distance of a star increases. A very distant star has virtually no parallax at all, so it can be treated as a motionless reference point against which the parallax of a nearby star could be measured. (Without some reference point, parallax is meaningless.)

However, astronomers had for centuries been trying to detect stellar parallaxes without success, although they had succeeded first with the parallax of the Moon, then of the Sun and the planets. Apparently, even the nearest stars had parallaxes so small as to make them difficult to measure.

Another trouble was that without knowing the parallaxes, one couldn't tell which star was near and which far. How, then, know which star to measure and which to use as a motionless reference point?

Astronomers made the general assumption that, all things being equal, a bright star is closer to Earth than is a dim star. Also, a star with high proper motion (a shift in position due to the star's own motion through space; a shift which is continuous — always in one direction and not cyclic, or back and forth, as parallactic shifts would be) was assumed nearer the Earth than one with a low proper motion. These assumptions would not necessarily hold true in every case, for a bright star might be more distant than a faint one, but be enough brighter, intrinsically, to make up for that. Again, a near star might have a very rapid apparent motion, but one which was in our line of sight so that it wouldn't show up. Nevertheless, these assumptions at least give astronomers a lead.

By the 1830s the time was ripe for a concerted attack on the problem. Three astronomers of three different nations tackled three different stars. Thomas Henderson (British) observed Alpha Centauri and Friedrich Wilhelm von Struve (German-born Russian) worked on Vega, the fourth brightest star in the sky. Both stars were not only bright but had pretty snappy proper motions. Friedrich Wilhelm Bessel (German) applied his efforts to 61 Cygni. This was a dim star but it had an unusually high proper motion. In each case, the star's position over at least a year was compared with that of a dim and presumably very far-off neighbour star.

Sure enough, each of the three stars being investigated shifted position slightly compared to its presumably distant neighbour. And so it happened (as it often does in science) that after centuries of failure, there were several almost simultaneous successes.

Bessel got in first, in 1839, and he gets the credit of being the first to measure the distance of a star. It turned out that 61 Cygni is 11 light-years distant. Henderson, later in 1839, reported Alpha Centauri to be a little over 4 light-years distant and Struve, in 1840, placed Vega at about 27 light-years distant.

No star has been found closer than those of the Alpha Centauri system.

Knowing the distance of Alpha Centauri, it is easy to calculate that Alpha Centauri A (the brighter of the pair) is almost exactly as bright as our Sun. Since its spectrum showed it to be at the same surface temperature, it is our Sun's twin — same diameter, same mass, same brightness, same everything, apparently.

As for Alpha Centauri B, if it were the same temperature

as Alpha Centauri A, then it would be just as luminous per unit area. To be only $1_3/._6$ as luminous as its companion, it must have $1_3/_6$ its area. The diameters of the two stars would be as the square roots of the respective areas and (assuming the two stars to be equally dense) the masses would be as the cube of the square roots of the respective areas.

It would then turn out that Alpha Centauri A would have a diameter 1·9 times that of Alpha Centauri B and a mass about 7 times that of Alpha Centauri B. (Actually, Alpha Centauri B is a trifle cooler than Alpha Centauri A, so that the comparison is not exactly as I've given it, but for the purposes of this essay, we needn't worry about the refinements.)

The two stars rotate in elliptical orbits about a common centre of gravity. The period of rotation is about 80 years. When the stars are closest, they are about a billion miles apart. When they are furthest they are 3·3 billion miles apart.

Now, then, suppose we try to duplicate (in imagination) the Alpha Centauri system here in our own Solar System. Since Alpha Centauri A is the twin of our Sun in every respect, let's suppose our Sun is Alpha Centauri A, but let's keep on referring to it, for convenience's sake, as the Sun.

Let's imagine Alpha Centauri B (which we will call simply Sun B) in orbit about the Sun. We can avoid unnecessary complications by making it exactly one half the diameter of the Sun and equally dense so that it is one eighth the Sun's mass. This may not be exactly the situation with respect to Alpha Centauri B, but it is a reasonable approximation.

Furthermore, let's suppose Sun B is travelling in a nearly circular orbit, in the same plane as the planets generally, and at the average distance of Alpha Centauri B from Alpha Centauri A (again a change in detail but not in essence). This would place it in orbit about 2,000,000,000 miles from the Sun. This is almost as though we have taken the planet

Uranus of our Solar System and replaced it with Alpha Centauri B.

All this would make Earth part of a multiple-star system very closely resembling that of Alpha Centauri. Now what would the heavens be like?

In some ways, our Solar System would be changed. Uranus, Neptune, and Pluto, as we know them, would be out. Their orbits would be tangled with Sun B. However, these planets were unknown in the pretelescopic era, so we can do without them as far as naked-eye observation is concerned.

But even Saturn, the outermost of the planets known to the ancients would be nearer to the Sun than to Sun B in the position I placed the latter. With the Sun, on top of that, having a gravitational field eight times as intense as that of Sun B, it should hang on to Saturn and the still closer planets with no trouble. (There might be interesting minor effects on the planetary orbits but I'm not astronomer enough, alas, to be able to calculate them.)

Sun B would behave like a new and very large 'planet' of the Sun. The Sun and Sun B would revolve about a centre of gravity which would be located in the asteroid belt. The motion of the Sun about this point once every eighty years would, however, not be detectable in pre-telescopic days, because the Sun would carry all the planets, including Earth, with it. Neither the Sun's distance nor Sun B's distance from Earth would be affected by that motion.

(After the invention of the telescope, the Sun's swing — with us in tow — would become noticeable through its reflection in the parallactic displacement of the nearer stars.)

But what would Sun B look like in our heavens?

Well, it would *not* look like a Sun. It would be a point of light like the other planets. A diameter of 430,000 miles at a distance of 2,000,000,000 miles would subtend an angle of about 45 seconds of arc. Sun B would appear to the naked eye to be just about the apparent size of the smaller but closer Jupiter.

To a naked-eye observer (such as the Greeks or Babylonians) Sun B would be one more point of light moving slowly against the stars. It would be moving more slowly than the others, making a complete circuit of the sky in about 80 years, as compared with 29½ years for Saturn and 12 for Jupiter. From this, the Greeks would — rightly — conclude that Sun B was further from Earth than was any other planet.

Of course, one thing would make Sun B very unusual and quite different from the other planets. It would be very bright. It would have an apparent magnitude of about −18. It would be only $1/3000$ as bright as the Sun, to be sure, but it would still be 150 times as bright as the full Moon. With Sun B in the night sky, Earth would be well-illuminated.

Another thing might be unusual about Sun B; not as a matter of inevitability, as with its brilliance, but as a matter of reasonable probability at least.

As a 'planet' of the Solar System, why should it not have satellites, as the other planets have? (Of course, its satellites would be revolving about a Sun and would really be planets, but let's not worry about having a consistent terminology.)

To be sure, since Sun B is much larger than the other planets, it could be expected to have a satellite much larger and more distant from itself than is true for any other planet.

It might, for instance, have a satellite the size of Uranus. (Why not? Uranus would be much smaller compared to Sun B, than Jupiter is compared to the Sun. If the Sun can have Jupiter in tow, then it is perfectly reasonable to allow Sun B to have a planet the size of Uranus.)

Uranus could be circling Sun B at a distance of 100,000,000 miles. (Again, why not? Jupiter, which is considerably smaller than Sun B and considerably closer to the competing gravity of the Sun nevertheless manages to hold on to satellites at a distance of 15,000,000 miles from itself. If Jupiter can manage that, Sun B can manage 100,000,000.)

If Uranus moved about Sun B in the plane of Earth's orbit, it would move first to one side of Sun B, then back and to the other side, then back and to the first side, and so on, indefinitely. Its maximum separation from Sun B would be about 3 degrees of arc. This is about 6 times the apparent diameter of the Sun or the Moon and such a separation could be easily seen with the naked eye.

But would Uranus itself be visible at that distance from us?

Well, right at the moment, *without* Sun B, Uranus *is* visible. It is 1,800,000,000 miles from the Sun (nearly as far as I have, in imagination, put Sun B) and it has a magnitude of 5·7 which makes it just visible as a very faint star.

But if Uranus were rotating about Sun B, it would be lit up not merely by the dim light of the distant Sun (the reflection of which is all we see Uranus by, actually) but also by the stronger rays of the much nearer Sun B.

The average magnitude of Uranus under these conditions would be 1·7. It wouldn't be as bright as the other planets, but it would be brighter than the North Star, for instance. The glare of the nearby Sun B might make Uranus harder to observe than the North Star, but it should still be clearly visible. (Sun B might have more than one satellite, too, but let's not complicate the picture. One satellite will do.)

The Greeks would thus be treated to the spectacle not only of an unusually and exceptionally brilliant point of light but also to another point of light (much dimmer) that oscillated back and forth as though caught in the grip of the brighter point.

Both factors, brilliance and a visible satellite, would be

completely unique. I have a theory that this would have made an interesting difference in Greek thinking, both on the mythological and the scientific level.

Mythology first (since Greek mythology is older than Greek science) and that involves the 'synodic period' of a planet. This is the interval between successive meetings of a planet and the Sun, in our sky. Jupiter and the Sun meet every 399 days; Saturn and the Sun every 378 days. Sun B and the Sun would meet in Earth's sky every 369 days. (This is just a measure of how frequently Earth in its revolution manages to get on the other side of the Sun from the planet in question.)

As the planet approaches the Sun it spends less and less time in the night sky and more and more time in the day sky. For ordinary planets this means it becomes less and less visible to the naked eye because it is lost in the Sun's glare during the day. Even the Moon looks washed out by day.

But Sun B would be different. Considering that it is 150 times as bright as the full Moon, it would be a clearly visible point of light even by day. Allowing the use of smoked glasses, it could be followed right up to the Sun.

Now the Greeks had a myth about how mankind learned the use of fire. At the time of creation, man was naked, shivering, and miserable; one of the weakest and most poorly endowed of the animal creation. The demigod Prometheus had pity on the new creature and stole fire from the Sun to give to mankind. With fire, man conquered night and winter and marauding beasts. He learned to smelt metals and developed civilisation.

But the anger of Zeus was kindled at this interference. Prometheus was taken to the very ends of the world (which, to the Greeks, were the Caucasus Mountains) and there chained to a rock. A vulture was sent there to tear at his liver every day, but it left him at night in order that his liver might miraculously grow back and be ready for the next day's torture.

There now. Doesn't all this fit in perfectly with the apparent behaviour of Sun B? Every year Sun B commits the crime of Prometheus. It can be seen in the daytime approaching the Sun, the only planet that can be seen to do this. It can only be planning to steal light from the Sun, and it obviously succeeds. After all, isn't that why it is so much brighter than all the other planets, why it is so much brighter even than the Moon?

Moreover, it brings this light to mankind, for when it is in the night sky, it illuminates the landscape into a dim kind of day.

But the planet is punished. It is cast out to the edge of the universe, further away than any other planet. There is even a vulture tearing at it, in the shape of its clearly visible satellite. While the planet was busy stealing fire from the Sun, no satellite was visible (because it was drowned out by the Sun's glare, of course). Once the planet was hurled to the edge of the universe, though, and became visible in the night sky, its satellite appeared. The satellite swoops towards the bright planet, tearing at it, then moves away to allow it to recover, then swoops in again, and so on in an eternal rhythm.

With all this in mind, isn't it just about inevitable that if Sun B were in our sky, it would be named Prometheus? Or that the satellite would now have the Latin name Vulturius.

Now I'm far too sober-minded and prosaic myself to think outlandish thoughts (as all of you know), but I wouldn't be surprised if some people reading this might not think the parallel is far too close to be accidental. Could it be that such a heavenly situation actually existed and suggested the myth in the first place?

Could it be that the human race originated on a planet

circling Alpha Centauri A? Could they have migrated to Earth about fifty thousand years ago, wiping out the primitive Neanderthals they found here and established a race of 'true men'? Could some disaster have destroyed their culture and forced them to build up a new one?

Is the Prometheus myth a dim memory of the distant past, when Alpha Centauri B lit up the skies? Was the Alpha Centauri system the original of the Atlantis myth?

No, I don't think so, but anyone who wants to use it in a science-fiction story is welcome to it. And anyone who wants to start a religious cult based on this notion probably can't be stopped but please — don't send me any of the literature — and *don't* say you read it here first.

And what effect would Sun B (or 'Prometheus') have had on Greek science?

Well, in the real world, there was a time when matters hung in the balance. The popular Greek theory of the universe, as developed by 300 B.C., put the Earth at the centre and let everything in the heavens revolve about it. The weight of Aristotle's philosophy was on the side of this theory.

About 280 B.C. Aristarchus of Samos suggested that only the Moon revolved about the Earth. The planets, including Earth itself, he said, revolved about the Sun, thus elaborating a heliocentric system. He even had some good notions concerning the relative sizes and distances of the Moon and the Sun.

For a while, the Aristarchean view seemed to have an outside chance despite the great prestige of Aristotle. However, about 150 B.C., Hipparchus of Nicaea worked out the mathematics of the geocentric system so thoroughly that the competition ended. About A.D. 150 Claudius Ptolemy put the final touches on the geocentric theory and no one questioned that the Earth was the centre of the universe for nearly 1,400 years thereafter.

But had Prometheus and Vulturius been in the sky, the Greeks would have had an example of one heavenly body, anyway, that clearly did not revolve primarily about the Earth. Vulturius would have revolved about Prometheus.

Aristarchus would undoubtedly have suggested Prometheus to be another sun with a planet circling it. The argument by analogy would, it seems to me, certainly have won out. Copernicus would have been anticipated.

Furthermore, the motions of Vulturius about Prometheus would have given a clear indication of the workings of gravity. The Aristotelian notion that gravity was confined to Earth alone and that heavenly bodies were immune to it would not have stood up.

Undoubtedly Newton too would have been anticipated by some two thousand years.

What would have happened next? Would Greek genius have decayed anyway? Would the Dark Ages still have intervened? Or would the world have had a two thousand year head start in science and would we now be masters of space? Or would we possibly be the non-survivors of a nuclear war fought in Roman times?

So that's how it goes. You start off checking on coloured shadows in a science-fiction story and end up wondering how different human history might have been (either for good or for evil) if only the Sun had had a companion star in its lonely voyage through eternity.

CHAPTER TWELVE

TWINKLE, TWINKLE, LITTLE STAR

It came as a great shock to me, in childhood days, to learn that our sun was something called a 'yellow dwarf' and that sophisticated people scorned it as a rather insignificant member of the Milky Way.

I had made the very natural assumption, prior to that, that stars were little things, and everything I had read confirmed the notion. There were innumerable fairy tales about the tiny stars, which (I gathered) must be the little children of the sun and moon, the brightly shining sun being the father and the dim, retiring moon the mother.

When I found that all those minute points of light were huge, glaring suns greater than our own, it not only upset the sanctity of the heavenly family for me, but it also offended me as a patriotic inhabitant of the solar system. Consequently, it was with grim relief that I eventually learned that not all stars were greater than the sun after all; that, in fact, a great many were smaller than the sun.

What's more, I found some of those small stars to be intensely fascinating; and in order to talk about them, I will begin Asimov-fashion at the other end of the stick, and consider the earth and the sun.

The earth does not really revolve about the sun. Both earth and sun (taken by themselves) revolve about a common centre of gravity. Naturally, the centre of gravity is closer to the more massive body and the degree of closeness is proportional to the ratio of masses of the two bodies.

Thus, the sun is 333,400 times as massive as the earth, and the centre of gravity should therefore be 333,400 times as close to the sun's centre as to the earth's centre. The distance between earth and sun, centre to centre, is about 92,870,000 miles; and dividing that by 333,400 gives us the figure 280. Therefore, the centre of gravity of the earth-sun system is 280 miles from the centre of the sun.

This means that as the earth moves around this centre of gravity in its annual revolution, the sun makes a small circle 280 miles in radius about the same centre, leaning always away from the earth. Of course, this trifling wobble is quite imperceptible from an observation point outside the solar system; say, from Alpha Centauri.

But what about the other planets? Each one of them revolves with the sun about a common centre of gravity. Some of the planets are both more massive than the earth and more distant from the sun, each of these factors working to move the centre of gravity farther from the sun's centre. To show you the result, I have worked out Table 32 (which, by the way, I have never seen in any astronomy text).

The radius of the sun is 432,200 miles, so the centre of gravity in every case but one lies below the sun's surface. The exception is Jupiter. The centre of gravity of the Jupiter-sun system is about 30,000 miles *above* the sun's surface (always in the direction of Jupiter, of course).

If the sun and Jupiter were all that existed in the solar system, an observer from Alpha Centauri, say, though not able to see Jupiter, might (in principle) be able to observe that the sun was making a tiny circle about something or other every twelve years. This 'something or other' could only be the centre of gravity of a system consisting of the sun and another body. If our observer had a rough idea of the mass of the sun, he could tell how distant the other body must be to impose a twelve-year revolution. From that distance, as compared with the radius of the circle the sun was making, he could deduce the mass of the other body. In this way, the observer on Alpha Centauri could discover the presence of Jupiter and work out its mass and its distance from the sun without ever actually seeing it.

Table 32

PLANET	DISTANCE (MILES) OF CENTRE OF GRAVITY OF SUN-PLANET SYSTEM FROM CENTRE OF SUN
Mercury	6
Venus	80
Earth	280
Mars	45
Jupiter	460,000
Saturn	250,000
Uranus	80,000
Neptune	140,000
Pluto	1,500 (?)

Actually, though, the wobble on the sun imposed by Jupiter is still too small to detect from Alpha Centauri (assuming their instruments to be no better than ours). What makes it worse is that Saturn, Uranus, and Neptune (the other planets can be ignored) impose wobbles on the sun, too, which complicate its motion.

But suppose that circling the sun were a body considerably more massive than Jupiter. The sun would then make a much larger circle and a much simpler one, for the effect of other revolving bodies would be swamped by this super-Jupiter. To be sure, this is not the case with the sun, but is it possible that it might be so for other stars?

Yes, indeed, it *is* possible.

In 1834 the German astronomer Friedrich Wilhelm Bessel concluded, from a long series of careful observations, that Sirius was moving across the sky in a wavy line. This could best be explained by supposing that the centre of gravity of Sirius and another body was moving in a straight line and that it was Sirius's revolution about the centre of gravity (in a period of some fifty years) that produced the waviness.

Sirius, however, is about two and a half times as massive as the sun, and for it to be pulled as far out of line as observation showed it to be, the companion body had to be much more massive than Jupiter. In fact, it turned out to be about one thousand times as massive as Jupiter, or just about as massive as our sun. If we call Sirius itself 'Sirius A', then this thousand-fold-Jupiter companion would be 'Sirius B'. (This use of letters has become a standard device for naming components of a multiple star system.)

Anything as massive as the sun ought to be a star rather than a planet and yet, try as he might, Bessel could see nothing in the neighbourhood of Sirius A where Sirius B ought to be. The seemingly natural conclusion was that Sirius B was a burned-out star, a blackened cinder that had used up its fuel. For a generation, astronomers spoke of Sirius's 'dark companion'.

In 1862, however, an American telescope-maker, Alvan Graham Clark, was testing a new eighteen-inch lens he had made. He turned it on Sirius to test the sharpness of the image it would produce, and, to his chagrin, found there was a flaw in his lens, for near Sirius was a sparkle of light that shouldn't be there. Fortunately, before going back to his grinding, he tried the lens on other stars, and the defect disappeared! Back to Sirius — and there was that sparkle of light again.

It couldn't be a defect; Clark had to be seeing a star. In fact, he was seeing Sirius's 'dark companion', which wasn't quite dark after all, for it was of the eighth magnitude. Allowing for its distance, however, it was at least dim, if not dark, for it was only $1/120$ as luminous as our sun — there was still that much of a dim glow amid its supposed ashes.

In the latter half of the nineteenth century, spectroscopy came into its own. Particular spectral lines could be produced only at certain temperatures, so that from the spectrum of a star its surface temperature could be deduced. In 1915 the American astronomer Walter Sydney Adams managed to get the spectrum of Sirius B and was amazed to discover that it was not a dimly glowing cinder at all, but had a surface rather hotter than that of the sun!

But if Sirius B was hotter than the sun, why was it only $1/120$ as bright as the sun? The only way out seemed to be to assume that it was much smaller than the sun and had, therefore, a smaller radiating surface. In fact, to account for both its temperature and its low total luminosity, it had to have a diameter of about 30,000 miles. Sirius B, although a star, was just about the size of the planet Uranus.

It was more dwarfish than any astronomer had conceived a star might be and it was white-hot, too. Consequently, Sirius B and all other stars of that type came to be called 'white dwarfs'.

But Bessel's observation of the mass of Sirius B was still valid. It was still just about as massive as the sun. To squeeze all that mass into the volume of Uranus meant that the average density of Sirius B had to be 38,000 kilograms per cubic centimetre, or about 580 tons per cubic inch.

Twenty years earlier, this consequence of Adams' discovery would have seemed so ridiculous that the entire chain of reasoning would have been thrown out of court and the very concept of judging stellar temperatures by spectral

lines would have come under serious doubt. By Adams'
time, however, the internal structure of the atom had been
worked out and it was known that virtually all the mass of
the atoms was concentrated in a tiny nucleus at the very
centre of the atom. If the atom could be broken down and
the central nuclei pushed together, the density of Sirius B —
and, in fact, densities millions of times greater still — became
conceivable.

Sirius B by no means represents a record either for the
smallness of a star or for its density. Van Maanen's Star
(named for its discoverer) has a diameter of only 6,048 miles,
so that it is smaller than the earth and not very much larger
than Mars. It is one seventh as massive as our sun (about 140
times as massive as Jupiter), and that is enough to make it
fifteen times as dense as Sirius B. A cubic inch of average
material from Van Maanen's Star would weigh 8,700 tons.

And even Van Maanen's Star isn't the smallest. In 1963
William J. Luyten of the University of Minnesota discovered
a white dwarf star with a diameter of about 1,000 miles —
only half that of the moon.[1]

Of course, the white dwarfs can't really give us much satis-
faction as 'little stars'. They may be dwarfs in volume but
they are sun-size in mass, and giants in density and in
intensity of gravitational fields. What about really little stars,
in mass and temperature as well as in volume?

These are hard to find. When we look at the sky, we are
automatically making a selection. We see all the large, bright
stars for hundreds of light-years in all directions, but the
dim stars we can barely see at all, even when they are fairly
close.

Judging by the stars we see, our sun, sure enough, is a
rather insignificant dwarf, but we can get a truer picture by
confining ourselves to our own immediate neighbourhood.

That is the only portion of space through which we can make a reasonably full census of stars, dim ones and all.

Thus, within five parsecs (16½ light-years) of ourselves, according to a compilation prepared by Peter Van de Kamp of Swarthmore College, there are thirty-nine stellar systems, including our own sun. Of these, eight include two visible components and two include three visible components, so that there are fifty-one individual stars altogether.

Of these, exactly three stars are considerably brighter than our sun and these we can call 'white giants'. They are listed in Table 33.

Table 33

STAR	DISTANCE (LIGHT-YEARS)	LUMINOSITY (SUN = 1)
Sirius A	8·6	23
Altair	15·7	8·3
Procyon A	11·0	6·4

There are then a dozen stars (*see* Table 34) that are as bright or nearly as bright as the sun. We can call these 'yellow stars' without making any invidious judgments as to whether they are dwarfs or not.

Of the remaining stars, (*see* Table 35) all of which are less than one twenty-fifth as luminous as the sun, four are white dwarfs.

This leaves thirty-two stars that are not only considerably dimmer than the sun, but considerably cooler, too, and therefore distinctly red in appearance. To be sure, there are cool red stars that are nevertheless much brighter in total luminosity than our sun because they are so gigantically voluminous. (This is the reverse of the white-dwarf situation.) These tremendous cool stars are 'red giants', and there are

Table 34

STAR	DISTANCE (LIGHT-YEARS)	LUMINOSITY (SUN = 1)
Alpha Centauri A	4·3	1·01
Sun	—	1·00
70 Ophiuchi A	16·4	0·40
Tau Ceti	11·2	0·33
Alpha Centauri B	4·3	0·30
Omicron$_2$ Eridani A	15·9	0·30
Epsilon Eridani	10·7	0·28
Epsilon Indi	11·2	0·13
70 Ophiuchi B	16·4	0·08
61 Cygni A	11·1	0·07
61 Cygni B	11·1	0·04
Groombridge 1618	14·1	0·04

none of these in the sun's vicinity — distant Betelgeuse and Antares are the best-known examples.

The cool, red, small stars are 'red dwarfs'. An example of this is the very nearest star to ourselves, the third and dimmest member of the Alpha Centauri system. It should be called Alpha Centauri C, but because of its nearness, it is more frequently called Proxima Centauri. It is only $1/_{23 \cdot 000}$ as bright as our sun and, despite its nearness, can be seen only with a good telescope.

Table 35

STAR	DISTANCE (LIGHT-YEARS)	LUMINOSITY (SUN = 1)
Sirius B	8·6	0·008
Omicron$_2$ Eridani B	15·9	0·004
Procyon B	11·0	0·0004
Van Maanen's Star	13·2	0·00016

To summarize, then, there are, in our vicinity: no red giants, three white giants, twelve yellow stars, four white dwarfs, and thirty-two red dwarfs. If we consider the immediate neighbourhood of the sun to be a typical one (and we have no reason to think otherwise), then well over half the stars in the heavens are red dwarfs and considerably dimmer than the sun. Indeed, our sun is among the top 10 per cent of the stars in luminosity — 'yellow dwarf' indeed!

The red-dwarf stars offer us something new. When I discussed the displacement of the sun by Jupiter at the beginning of the article, I pointed out that the displacement would be larger, and therefore possible to observe from other stars, if Jupiter were considerably larger.

An alternative would be to have the sun considerably less massive. It is not the absolute mass of either component, but the ratio of the masses that counts. Thus, the Jupiter-sun ratio is 1:1000, which leads to an indetectable displacement. The mass ratio of the two components of the Sirius system, however, is 1:2·5, and that is easily detectable.

If a star were, say, half the mass of the sun, and if it were circled by a body eight times the mass of Jupiter, the mass ratio would be about 1:60. The displacement would not be as readily noticeable as in the case of Sirius, but it would be detectable.

Exactly such a displacement was detected in 1943 at Sproul Observatory in Swarthmore College, in connection with 61 Cygni. From unevennesses in the motion of one of the major components, a third component, 61 Cygni C, was deduced as existing; a body with a mass $1/125$ of our sun or only eight times that of Jupiter. In 1960 similar displacements were discovered for the star Lalande 21185 at Sproul Observatory. It, too, had a planet eight times the mass of Jupiter.

And in 1963, the same observatory announced a third planet outside the solar system. The star involved is Barnard's Star.

This star was discovered in 1916 by the American astronomer Edward Emerson Barnard, and it turned out to be an unusual star indeed. In the first place it is the second nearest star to ourselves, being only 6·1 light-years distant. (The three stars of the Alpha Centauri system, considered as a unit, are the nearest, at 4·3 light-years; Lalande 21185 at 7·9 light-years is third nearest. Next is Wolf 359 and then the two stars of the Sirius system — at 8·0 and 8·6 light-years respectively.)

Barnard's Star has the most rapid proper motion known, partly because it is so close. It moves 10·3 seconds of arc a year. This isn't much, really, for in the forty-seven years since its discovery it has only moved a little over 8 minutes of arc (or about one quarter the apparent width of the moon) across the sky. For a 'fixed' star, however, that's a tremendously rapid movement; so rapid, in fact, that the star is sometimes called 'Barnard's Runaway Star' or even 'Barnard's Arrow'.

Barnard's Star is a red dwarf with about one fifth the mass of the sun and less than $1/2500$ the luminosity of the sun (though it is nine times as luminous as Proxima Centauri).

The planet displacing Barnard's Star is Barnard's Star B and it is the smallest of the three invisible bodies yet discovered. It is about $1/700$ the mass of the sun and hence roughly 1·2 times the mass of Jupiter. Put another way, it is about five hundred times the mass of the earth. If it possesses the over-all density of Jupiter, it would make a planetary body about 100,000 miles in diameter.

All this has considerable significance. Astronomers have about decided from purely theoretical considerations that most stars have planets. Now we find that in our immediate neighbourhood at least three stars have at least one planet apiece. Considering that we can only detect super-Jovian planets, this is a remarkable record. Our sun has one planet of Jovian size and eight sub-Jovians. It is reasonable to suppose that any other star with a Jovian planet has a family of sub-Jovians also. And indeed, there ought to be a number of stars with sub-Jovian planets only.

In short, on the basis of these planetary discoveries, it would seem quite likely that nearly every star has planets.

A generation ago, when it was believed that solar systems arose through collisions or near-collisions of stars, it was felt that a planetary family was excessively rare. Now we might conclude that the reverse is true; it is the truly lone star, the one without companion stars or planets, that is the really rare phenomenon.

And yet the red dwarfs aren't quite as little as they seem to be from their luminosity. Even the smallest red dwarf, Proxima Centauri, is not less than one tenth the mass of the sun. In fact, stellar masses are quite uniform; much more uniform than stellar volumes, densities, or luminosities. Virtually all stars range in mass from not less than one tenth of the sun to not more than ten times the sun, a stretch of but two orders of magnitude.

There is good reason for this. As mass increases, the pressure and temperature at the centre of the body also increases and the amount of radiation produced varies as the fourth power of the temperature. Increase the temperature ten times, in other words, and luminosity increases ten thousand times.

Stars that are more than ten times the mass of the sun are therefore unstable, for the pressures associated with their vastly intense radiations blow them apart in short order. On the other hand, stars with less than one tenth the mass of the sun do not have an internal temperature and pressure high enough to start a self-sustaining nuclear reaction.

The upper limit is fairly sharp. Too-massive stars, except in very rare cases, blow up and actually don't exist. Too-light stars merely don't shine and can't be seen, so that the lower limit is an arbitrary one. The light bodies may exist even if they can't be seen.

Below the smallest luminous stars are, indeed, the non-luminous planets. In our own solar system, we have bodies up to the size of Jupiter, which is perhaps $1/100$ the mass of the feebly glowing Proxima Centauri. A body such as 61 Cygni C would have a mass one twelfth that of Proxima Centauri. Undoubtedly there must be bodies closing that remaining gap in mass.

Jupiter, large as it is for a planet, develops insufficient heat at its centre to lend significant warmth to its surface. Whatever warmth exists on Jupiter's surface derives from solar radiation.[2] The same may be true for 61 Cygni C.

However, as we consider planets larger still, there must come a point where the internal heat, while not great enough to start runaway nuclear reactions, is great enough to keep the surface warm, perhaps warm enough to allow water to remain eternally in the liquid form.

We might call this a super-planet but, after all, it is radiating energy in the infrared. Such a body would not glow visibly, but if our eyes were sensitive to infrared we might see them as very dim stars. They might, therefore, be more fairly called 'sub-stars' than super-planets.

Harlow Shapley, emeritus director of Harvard College Observatory, thinks it possible that such sub-stars are very common in space, and that they might even be the abode of life. To be sure, a sub-star with an earth-like density would have a diameter of about 150,000 miles and a surface gravity about eighteen times earth-normal. To life developing in the oceans, however, gravity is of no importance.

Is it possible that such a sub-star (with, perhaps, a load of life) might come rolling close enough to the solar system, some day, to attract exploring parties?

We can't be certain it won't happen. In the case of luminous stars, we can detect invaders from afar, and we can be certain that none will be coming this way for millions of years. A sub-star, however, could sneak up on us unobserved; we'd never know it was approaching. It might be right on top of us — say, within fifteen billion miles of the sun — before we detected its presence through its reflected light and through its gravitational perturbations on the outer planets.

Then at last mankind might go out to see for themselves what a little star was like and set to rest that generations-long plaintive chant of childhood. 'How I wonder what you are!'

Only — it won't be twinkling.

CHAPTER THIRTEEN

HEAVEN ON EARTH

The nicest thing about writing these essays is the constant mental exercise it gives me. Unceasingly, I must keep my eyes and ears open for anything that will spark something that will, in my opinion, be of interest to the reader.

For instance, a letter arrived today, asking about the duodecimal system, where one counts by twelves rather than by tens, and this set up a mental chain reaction that ended in astronomy and, what's more, gave me a notion which, as far as I know, is original with me. Here's how it happened.

My first thought was that, after all, the duodecimal system *is* used in odd corners. For instance, we say that 12 objects make 1 dozen and 12 dozen make 1 gross. However, as far as I know, 12 has never been used as the base for a number system, except by mathematicians in play.

A number which has, on the other hand, been used as the base for a formal positional notation is 60. The ancient Babylonians used 10 as a base just as we do, but frequently used 60 as an alternate base. In a number based on 60, what we call the units column would contain any number from 1 to 59, while what we call the tens column would be the 'sixties' column, and our hundreds column (ten times ten) would be the 'thirty-six hundred' column (sixty times sixty).

Thus, when we write a number, 123, what it really stands for is $(1 \times 10^2) + (2 \times 10^1) + (3 \times 10^0)$. And since 10^2 equals

100, 10^1 equals 10 and 10^0 equals 1, the total is $100+20+3$ or, as aforesaid, 123.

But if the Babylonians wrote the equivalent of 123, using 60 as the base, it would mean $(1\times60^2)+(2\times60^1)+(3\times60^0)$. And since 60^2 equals 3600, 60^1 equals 60 and 60^0 equals 1, this works out to $3600+120+3$, or, 3,723 by our decimal notation. Using a positional notation with the base 60 is a 'sexagesimal notation' from the Latin word for sixtieth.

As the word 'sixtieth' suggests, the sexagesimal notation can be carried into fractions too.

Our own decimal notation will allow us to use a figure such as 0·156, where what is really meant is $0+{}^1/_{10}+{}^5/_{100}+{}^6/_{1000}$. The denominators, you see, go up the scale in multiples of 10. In the sexagesimal scale, the denominators would go up the scale in multiples of 60 and 0.156 would represent $0+{}^1/_{60}+{}^5/_{3600}+{}^6/_{216,000}$, since 3600 equals 60×69, 216,000 equals $60\times60\times60$, and so on.

Those of you who know all about exponential notation will no doubt be smugly aware that ${}^1/_{10}$ can be written 10^{-1}, ${}^1/_{100}$ can be written 10^{-2} and so on, while ${}^1/_{60}$ can be written 60^{-1}, ${}^1/_{3600}$ can be written 60^{-2} and so on. Consequently, a full number expressed in sexagesimal notation would be something like this: (15) (45) (2). (17) (25) (59), or $(15\times60^2)+(45\times60^1)+(2\times60^0)+(17\times60^{-1})+(25\times60^{-2})+(59\times60^{-3})$, and if you want to amuse yourself by working out the equivalent in ordinary decimal notation, please do. As for me, I'm chickening out right now.

All this would be of purely academic interest, if it weren't for the fact that we still utilize sexagesimal notation in at least two important ways, which date back to the Greeks.

The Greeks had a tendency to pick up the number 60 from the Babylonians as a base, where computations were complicated, since so many numbers go evenly into 60 that fractions are avoided as often as possible (and who wouldn't avoid fractions as often as possible?).

One theory, for instance, is that the Greeks divided the radius of a circle into 60 equal parts so that in dealing with half a radius, or a third, or a fourth, or a fifth, or a sixth, or a tenth (and so on) they could always express it as a whole number of sixtieths. Then, since in ancient days the value of π (pi) was often set equal to a rough and ready 3, and since the length of the circumference of a circle is equal to twice π times the radius, the length of that circumference is equal to 6 times the radius or to 360 sixtieths of a radius. Thus (perhaps) began the custom of dividing a circle into 360 equal parts.

Another possible reason for doing so rests with the fact that the sun completes its circuit of the stars in a little over 365 days, so that in each day it moves about $1/365$ of the way around the sky. Well, the ancients weren't going to quibble about a few days here and there and 360 is so much easier to work with that they divided the circuit of the sky into that many divisions and considered the sun as travelling through one of those parts (well, just about) each day.

A 360th of a circle is called a 'degree' from Latin words meaning 'step down'. If the sun is viewed as travelling down a long circular stairway, it takes one step down (well, just about) each day.

Each degree, if we stick with the sexagesimal system, can be divided into 60 smaller parts and each of those smaller parts into 60 still smaller parts and so on. The first division was called in Latin *pars minuta prima* (first small part) and the second was called *pars minuta secunda* (second small part), which have been shortened in English to minutes and seconds respectively.

We symbolize the degree by a little circle (naturally), the minute by a single stroke, and the second by a double stroke, so that when we say that the latitude of a particular spot on earth is 39° 17′ 42″, we are saying that its distance from the equator is 39 degrees plus $17/60$ of a degree plus $42/3600$ of a degree, and isn't that the sexagesimal system?

The second place where sexagesimals are still used is in measuring time (which was originally based on the movements of heavenly bodies). Thus we divide the hour into minutes and seconds and when we speak of a duration of 1 hour, 44 minutes, and 20 seconds, we are speaking of a duration of 1 hour plus $44/60$ of an hour plus $20/3600$ of an hour.

You can carry the system further than the second and, in the Middle Ages, Arabic astronomers often did. There is a record of one who divided one sexagesimal fraction into another and carried out the quotient to ten sexagesimal places, which is the equivalent of 17 decimal places.

Now let's take sexagesimal fractions for granted, and let's consider next the value of breaking up circumferences of circles into a fixed number of pieces. And, in particular, consider the circle of the ecliptic along which the sun, moon, and planets trace their path in the sky.

After all, how *does* one go about measuring a distance along the sky? One can't very well reach up with a tape measure. Instead the system, essentially, is to draw imaginary lines from the two ends of the distance traversed along the ecliptic (or along any other circular arc, actually) to the centre of the circle, where we can imagine our eye to be, and to measure the angle made by those two lines.

The value of this system is hard to explain without a diagram, but I shall try to do so, with my usual dauntless bravery (though you're welcome to draw one as I go along, just in case I turn out to be hopelessly confusing).

Suppose you have a circle with a diameter of 115 feet, and

another circle drawn about the same centre with a diameter of 230 feet, and still another drawn about the same centre with a diameter of 345 feet. (These are 'concentric circles' and would look like a target.)

The circumference of the innermost circle would be about 360 feet, that of the middle one 720 feet and that of the outermost 1,080 feet.

Now mark off $1/360$ of the innermost circle's circumference, a length of arc 1 foot long, and from the two ends of the arc draw lines to the centre. Since $1/360$ of the circumference is 1 degree, the angle formed at the centre may be called 1 degree also (particularly since 360 such arcs will fill the entire circumference and 360 such central angles will consequently fill the entire space about the centre).

If the 1-degree angle is now extended outwards so that the arms cut across the two outer circles, they will subtend a 2-foot arc of the middle circle and a 3-foot circle of the outer one. The arms diverge just enough to match the expanding circumference. The lengths of the arc will be different, but the fraction of the circle subtended will be the same. A 1-degree angle with vertex at the centre of a circle will subtend a 1-degree arc of the circumference of any circle, regardless of its diameter, whether it is the circle bounding a proton or bounding the Universe (if we assume a Euclidean geometry, I quickly add). The same is analogously true for an angle of any size.

Suppose your eye was at the centre of a circle that had two marks upon it. The two marks are separated by $\frac{1}{6}$ the circumference of the circle, that is by $360/6$ or 60 degrees of arc. If you imagine a line drawn from the two marks to your eye, the lines will form an angle of 60 degrees. If you look first at one mark, then at the other you will have to swivel your eyes through an angle of 60 degrees.

And it wouldn't matter, you see, whether the circle was a

mile from your eye or a trillion miles. If the two marks were ⅙ of the circumference apart, they would be 60 degrees apart, regardless of distance. How nice to use such a measure, then, when you haven't the faintest idea of how far away the circle is.

So, since through most of man's history astronomers had no notion of the distance of the heavenly objects in the sky, angular measure was just the thing.

And if you think it isn't, try making use of linear measure. The average person, asked to estimate the diameter of the full moon *in appearance,* almost instinctively makes use of linear measure. He is liable to reply, judiciously, 'Oh, about a foot.'

But as soon as he makes use of linear measure, he is setting a specific distance, whether he knows it or not. For an object a foot across to look as large as the full moon, it would have to be 36 yards away. I doubt that anyone who judges the moon to be a foot wide will also judge it to be no more than 36 yards distant.

If we stick to angular measure and say that the average width of the full moon is 31′ (minutes), we are making no judgments as to distance and are safe.

But if we're going to insist on using angular measure, with which the general population is unacquainted, it becomes necessary to find some way of making it clear to everyone. The most common way of doing this, and to picture the moon's size, for instance, is to take some common circle with which we are all acquainted and calculate the distance at which it must be held to look as large as the moon.

One such circle is that of the twenty-five-cent piece. Its diameter is about 0·96 inches and we won't be far off if we consider it just an inch in diameter. If a quarter is held 9 feet from the eye, it will subtend an arc of 31 minutes. That means it will look just as large as the full moon does, and, if it is held at that distance between your eye and the full moon, it will just cover it.

Now if you've never thought of this, you will undoubtedly be surprised that a quarter at 9 feet (which you must imagine will look quite small) can overlap the full moon (which you probably think of as quite large). To which I can only say: Try the experiment!

Well, this sort of thing will do for the sun and the moon but these, after all, are, of all the heavenly bodies, the largest in appearance. In fact, they're the only ones (barring an occasional comet) that show a visible disc. All other objects are measured in fractions of a minute or even fractions of a second.

It is easy enough to continue the principle of comparison by saying that a particular planet or star has the apparent diameter of a quarter held at a distance of a mile or ten miles or a hundred miles and this is, in fact, what is generally done. But of what use is that? You can't see a quarter at all, at such distances, and you can't picture its size. You're just substituting one unvisionable measure for another.

There must be some better way of doing it.

And at this point in my thoughts, I had my original (I hope) idea.

Suppose that the earth were exactly the size it is but were a huge hollow, smooth, transparent sphere. And suppose you were viewing the skies not from earth's surface, but with your eye precisely at earth's centre. You would then see all the heavenly objects projected on to the sphere of the earth.

In effect, it would be as though you were using the entire globe of the earth as a background on which to paint a replica of the celestial sphere.

The value of this is that the terrestrial globe is the one sphere upon which we can easily picture angular measurement, since we have all learned about latitude and longitude which *are* angular measurements. On the earth's surface,

1 degree is equal to 69 miles (with minor variations, which we can ignore, because of the fact that the earth is not a perfect sphere). Consequently, 1 minute, which is equal to $1/60$ of a degree, is equal to 1·15 miles or 6,060 feet, and 1 second, which is equal to $1/60$ of a minute, is equal to 101 feet.

You see, then, that if we know the apparent angular diameter of a heavenly body, we know exactly what its diameter would be if it were drawn on the earth's surface to scale.

The moon, for instance, with an average diameter of 31 minutes by angular measure, would be drawn with a diameter of 36 miles, if painted to scale on the earth's surface. It would neatly cover all of Greater New York or the space between Boston and Worcester.

Your first impulse may be a 'WHAT!' but this is not really as large as it seems. Remember, you are really viewing this scale model from the centre of the earth, four thousand miles from the surface, and just ask yourself how large Greater New York would seem, seen from a distance of 4,000 miles. Or look at a globe of the earth, if you have one and picture a circle with a diameter stretching from Boston to Worcester and you will see that it is small indeed compared to the whole surface of the earth, just as the moon itself is small indeed compared to the whole surface of the sky. (Actually, it would take the area of 490,000 bodies the size of the moon to fill the entire sky, and 490,000 bodies the size of our painted moon to fill all of the earth's surface.)

But at least this shows the magnifying effect of the device I am proposing, and it comes in particularly handy where bodies smaller in appearance than the sun or the moon are concerned just at the point where the quarter-at-a-distance-of-so-many-miles notion breaks down.

For instance, in Table 36, I present the maximum angular diameters of the various planets as seen at the time of their closest approach to earth, together with their linear diameter to scale if drawn on earth's surface.

I omit Pluto because its angular diameter is not well known. However, if we assume that planet to be about the size of Mars then at its furthest point in its orbit, it will still have an angular diameter of 0·2 seconds and can be presented as a circle 20 feet in diameter.

Table 36

PLANETS TO SCALE

PLANET	ANGULAR DIAMETER (SECONDS)	LINEAR DIAMETER (FEET)
Mercury	12·7	1,280
Venus	64·5	6,510
Mars	25·1	2,540
Jupiter	50·0	5,050
Saturn	20·6	2,080
Uranus	4·2	425
Neptune	2·4	240

Each planet could have its satellites drawn to scale with great convenience. For instance, the four large satellites of Jupiter would be circles ranging from 110 to 185 feet in diameter, set at distances of 3 to 14 miles from Jupiter. The entire Jovian system to the orbit of its outermost satellite (Jupiter IX, a circle about 5 inches in diameter) would cover a circle about 350 miles in diameter.

The real interest in such a setup, however, would be the stars. The stars, like the planets, do not have a visible disc to the eyes. Unlike the planets, however, they do not have a visible disc even to the largest telescope. The planets (all but Pluto) can be blown up to discs even by moderate-sized telescopes; not so the stars.

By indirect methods the apparent angular diameter of some stars has been determined. For instance, the largest angular diameter of any star is probably that of Betelgeuse, which is 0·047 seconds. Even the huge 200-inch telescope cannot magnify that diameter more than a thousandfold, and under such magnification the largest star is still less than 1 minute of arc in appearance and is therefore no more of a disc to the 200-incher than Jupiter is to the unaided eye. And of course, most stars are far smaller in appearance than is giant Betelgeuse. (Even stars that are in actuality larger than Betelgeuse are so far away as to appear smaller.)

But on my earth scale, Betelgeuse with an apparent diameter of 0·047 seconds of arc would be represented by a circle about 4·7 feet in diameter. (Compare that with the 20 feet of even distantest Pluto.)

However, it's no use trying to get actual figures on angular diameters because these have been measured for very few stars. Instead, let's make the assumption that all the stars have the same intrinsic brightness the sun has. (This is not so, of course, but the sun is an average star, and so the assumption won't radically change the appearance of the universe.)

Now then, area for area, the sun (or any star) remains at constant brightness to the eye regardless of distance. If the sun were moved out to twice its present distance, its apparent brightness would decrease by four times but so would its apparent surface area. What we could see of its area would be just as bright as it ever was; there would be less of it, that's all.

The same is true the other way, too. Mercury, at its closest approach to the sun, sees a sun that is no brighter per square second than ours is, but it sees one with ten times as many square seconds as ours has, so that Mercury's sun is ten times as bright as ours in total.

Well, then, if all the stars were as luminous as the sun, then the apparent area would be directly proportional to the apparent brightness. We know the magnitude of the sun (−26·72) as well as the magnitudes of the various stars, and that gives us our scale of comparative brightness, from which we can work out a scale of comparative areas and, therefore, comparative diameters. Furthermore, since we know the angular measure of the sun, we can use the comparative diameters to calculate the comparative angular measures which, of course, we can convert to linear diameters (to scale) on the earth.

But never mind the details (you've probably skipped the previous paragraph already), I'll give you the results in Table 36.

(The fact that Betelgeuse has an apparent diameter of 0·047 and yet is no brighter than Altair is due to the fact that Betelgeuse, a red giant, has a lower temperature than the sun and is much dimmer per unit area in consequence. Remember that Table 37 is based on the assumption that all stars are as luminous as the sun.)

So you see what happens once we leave the Solar System. Within that system, we have bodies that must be drawn to scale in yards and miles. Outside the system, we deal with bodies which, to scale, range in mere inches.

If you imagine such small patches of earth's surface, as seen from the earth's centre, I think you will get a new vision as to how small the stars are in appearance and why telescopes cannot make visible discs of them.

The total number of stars visible to the naked eye is about 6,000, of which two thirds are dim stars of 5th and 6th magnitude. We might then picture the earth as spotted with 6,000 stars, most of them being about an inch in diameter. There would be only a very occasional larger one; only 20, all told, that would be as much as 6 inches in diameter.

The average distance between two stars on earth's surface

Table 37

STARS TO SCALE

MAGNITUDE OF STAR	ANGULAR DIAMETER (SECONDS)	LINEAR DIAMETER (INCHES)
−1 (e.g. Sirius)	0·014	17·0
0 (e.g. Rigel)	0·0086	10·5
1 (e.g. Altair)	0·0055	6·7
2 (e.g. Polaris)	0·0035	4·25
3	0·0022	2·67
4	0·0014	1·70
5	0·00086	1·05
6	60·00055	0·67

would be 180 miles. There would be one or, at most, two stars in New York State, and one hundred stars, more or less, within the territory of the United States (including Alaska).

The sky, you see, is quite uncrowded, regardless of its appearance.

Of course, these are only the visible stars. Through a telescope, myriads of stars too faint to be seen by the naked eye can be made out and the 200-inch telescope can photograph stars as dim as the 22nd magnitude.

A star of magnitude 22, drawn on the earth to scale, would be a mere 0·0004 inches in diameter, or about the size of a bacterium. (Seeing a shining bacterium on earth's surface from a vantage point at earth's centre, 4,000 miles down, is a dramatic indication of the power of the modern telescope.)

The number of individual stars visible down to this magnitude would be roughly two billion. (There are, of course, at least a hundred billion stars in our Galaxy, but almost all of them are located in the Galactic nucleus which is completely hidden from our sight by dust clouds. The two billion we do see are just the scattering in our own neighbourhood of the spiral arms.)

Drawn to scale on the earth, this means that among the 6,000 circles we have already drawn (mostly an inch in diameter) we must place a powdering of two billion more dots, a small proportion of which are still large enough to see, but most of which are microscopic in size.

The average distance between stars even after this mighty powdering would still be, on the earth-scale, 1,700 feet.

This answers a question I, for one, have asked myself in the past. Once a person looks at a photograph showing the myriad stars visible to a large telescope, he can't help wondering how it is possible to see beyond all that talcum powder and observe the outer galaxies.

Well, you see, despite the vast numbers of stars, the clear space between them is still comparatively huge. In fact, it has been estimated that all the starlight that reaches us is equivalent to the light of 1,100 stars of magnitude 1. This means that if all the stars that can be seen were massed together, they would fill a circle (on earth-scale) that would be 18·5 feet in diameter.

From this we can conclude that all the stars combined do not cover up as much of the sky as the planet Pluto. As a matter of fact, the moon, all by itself, obscures nearly 300 times as much of the sky as do all other nighttime heavenly bodies, planets, satellites, planetoids, stars, put together.

There would be no trouble whatever in viewing the spaces outside our Galaxy if it weren't for the dust clouds. These are really the only obstacle, and they can't be removed even if we set up a telescope in space.

What a pity the universe couldn't really be projected on earth's surface temporarily — just long enough to send out the Walrus's seven maids with seven mops with strict orders to give the universe a thorough dusting.

How happy astronomers would then be![1]

CHAPTER FOURTEEN

THE FLICKERING YARDSTICK

Every once in a while, astronomical opinion concerning the size of the Universe changes suddenly — invariably for the larger. The last time this happened, the responsibility could be placed directly at the door of a wartime blackout.

As late as the turn of the century, astronomers had only the foggiest notion of the size of the Universe, as a matter of fact. The best estimate of the time was made by a Dutch astronomer named Jacobus Cornelis Kapteyn. Beginning in 1906, he supervised a survey of the Milky Way. He would photograph small sections of it and count the stars of the various magnitudes. Assuming them to be average-sized stars, he calculated the distance they would have to be in order to show up as dimly as they did.

He ended up with the concept of the Galaxy as a lens-shaped object (something which had been more or less generally accepted since the days of William Herschel, a century earlier). The Milky Way is simply the cloudy haze formed by the millions of distant stars we see when we look through the Galactic lens the long way. Kapteyn estimated that the Galaxy was 23,000 light-years in extreme diameter and 6,000 light-years thick. And as far as he, or anyone, could tell at the time, nothing existed outside the Galaxy.

He also decided that the Solar System was located quite near the centre of the Galaxy, by the following line of reasoning. First, the Milky Way cut the heavens into approximately equal halves, so we must be on the median plane of the lens. If we were much above or below the median plane, the Milky Way would be crowded into one half of the sky.

Secondly, the Milky Way was about equally bright all the way around. If we were towards one end or the other of the lens, the Milky Way in the direction of the farther end would be thicker and brighter than the section in the direction of the nearer end.

In short, the Sun is in the centre of the Galaxy, more or less, because the heavens, as seen from Earth, are symmetrical, and there you are.

But there was one characteristic of the heavens which showed a disturbing asymmetry. Present in the sky are a number of 'globular clusters'. These are collections of stars packed rather tightly into a more or less spherical shape. Each globular cluster contains anywhere from a hundred thousand stars to a few million and about two hundred of them exist in our Galaxy.

Well, there's no reason why such clusters shouldn't be evenly distributed in the Galaxy, and if we were at the centre, they should be spread evenly all over the sky — but they're not. A large part of them seem to be crowded together in one small part of the sky, that part covered by the constellations of Sagittarius and Scorpio.

It's the sort of odd fact that bothers astronomers and often proves the gateway to important new views of the Universe.

The way to a solution of the problem, and to the new view of the Universe, lay through a consideration of a certain kind of variable star; a star that is which is constantly varying in brightness; a star which flickers, if you like.

There are a number of different kinds of variable stars, differentiated from one another by the exact pattern of light variation. Some stars flicker for outside reasons; usually because they are eclipsed by a dim companion which gets in the way of our line of sight. The star Algol in the constellation Perseus has a dim companion which gets in our way every 69 hours. During that time of eclipse, Algol loses two-thirds of its light (it is not a total eclipse) in a matter of a couple of hours and regains it as quickly.

More interesting are stars which *really* vary in brightness because of changes in their internal constitution. Some explode with greater or lesser force; some vary all over the lot in irregular fashion for mysterious reasons; and some vary in very regular fashion for equally mysterious reasons.

One of the brighter and more noticeable examples of the last group is a star called Delta Cephei, in the constellation Cepheus. It brightens, dims, brightens, dims within a period of 5·37 days. From its dimmest point it brightens steadily for about two days, reaching a peak brightness that is just double its brightness at dim point. It then spends about three and a third days fading off to its dim point again. The brightening is distinctly more rapid than the dimming.

From its spectrum, it would seem that Delta Cephei is a pulsating star. That is, it expands and contracts. If it remained the same temperature during this pulsation, it would be easy to understand that it was brightest at peak size and dimmest at least size. However, it also changes temperature and is hottest at peak brightness and coolest at the dim point. The trouble is that the peak temperature and peak brightness come, not at maximum size, but when it is expanding and halfway towards peak size. The lowest temperature and dimmest point comes when it is contracting and halfway towards minimum size. This means that Delta Cephei ends up being about the same size at the peak of brightness and in the trough of dimness. In the first case, though, it is in the process of expansion; in the latter, in the process of contraction.

Why the regular but non-synchronous pulsation in size and temperature? That part is still a mystery.

There is enough that is characteristic of Delta Cephei in all this to make astronomers realise when they found other stars behaving in like fashion, that all must belong to a group of structurally similar stars, which they called 'Cepheid variables' in honour of the first of the group.

Cepheids vary among themselves in the length of their period. Some periods are as short as one day, some as long as 45 days, with examples all the way through the range. The Cepheids closest to us have periods of about a week.

The brightest and closest Cepheid is none other than the North Star. It has a period of 4 days, but during that time its flicker causes it to vary in brightness by only about 10 per cent, so it's not surprising no non-astronomer notices it, and that astronomers themselves paid more attention to the somewhat dimmer but more drastically changeable Delta Cephei.

There are a number of stars with Cepheid-like variation curves that are to be found in the globular clusters. Their main distinction from the ordinary Cepheids nearer us, is that they are extremely short-period. The longest period among these is about a day, and periods as short as an hour and a half are known. These were first called cluster Cepheids while ordinary Cepheids were called classical Cepheids. However, cluster Cepheids turned out to be a misnomer, because such stars were found with increasing frequency outside clusters, also.

The cluster Cepheids are now usually called by the name of the best-studied example (just as the Cepheids themselves are). The best-studied example is a star called RR Lyrae, so the cluster Cepheids are called RR Lyrae variables.

Now none of this seemed to have any connection with the

size of the Universe until 1912 when Miss Henrietta Leavitt, studying the Small Magellanic Cloud, came across a couple of dozen Cepheids in them.

(The Large and the Small Magellanic Clouds are two foggy patches that look like detached remnants of the Milky Way. They are visible in the Southern Hemisphere and were first sighted by Europeans during the round-the-world voyage of Ferdinand Magellan and his crew back in 1520 — whence the name.)

The Magellanic Clouds can be broken up into stars by a good telescope and it is only because they are a long way distant from us that these stars fade into an undifferentiated foggy patch. Because the clouds are so far distant from us, all the stars in one cloud or the other may be considered about the same distance from us. Whether a particular star is at the near edge of the cloud or the far edge makes little difference. (This is like making the equally true statement that every man in the state of Washington is roughly the same distance, i.e. 3,000 miles, from an individual in Boston.)

This also means that when one star in the Small Magellanic Cloud appears twice as bright as another star, it *is* twice as bright. There is no distance difference to confuse the issue.

Well, when Miss Leavitt recorded the brightness and the period of variation of the Cepheids in the Small Magellanic Cloud, she found a smooth relationship. The brighter the Cepheid, the longer the period. She prepared a graph correlating the two and this is called the 'period-luminosity curve'.

Such a curve could not have been discovered from the Cepheids near us, just because of the confusing distance difference. For instance, Delta Cephei is more luminous than the North Star and therefore has a longer period. But the North Star is considerably closer to us than is Delta Cephei, so that the North Star *seems* brighter to us. For that reason, the longer period *seems* to go with the dimmer star.

Of course, if we knew the actual distances of the North Star and of Delta Cephei, we could straighten the matter out, but at the time the distances were not known.

Once the period-luminosity curve was established, astronomers promptly made the assumption that it held for all Cepheids, and were then able to make a scale model of the Universe. That is, if astronomers spotted two Cepheids with equal periods, they could assume they were also equal in actual luminosity. If Cepheid A seemed only one fourth as bright as Cepheid B, it could only be because Cepheid A was twice as distant from us as was Cepheid B. (Brightness varies inversely as the square of the distance.) Cepheids of different period could be placed, relatively to us, with only slightly more trouble.

With all the Cepheids in relative place, astronomers would need to know the actual distance in light-years to any one of them, in order to know the actual distance to all the rest.

There was only one trouble here. The sure method of determining the distance of a star was to measure its parallax. At distances of more than 100 light-years, however, the parallax becomes too small to measure. And unfortunately even the nearest Cepheid, the North Star, is several times that distance from us.

Astronomers were forced into long-drawn-out, complex, statistical analyses of medium-distant (not globular) star clusters. In this way, they determined the actual distance of some of those clusters, including the Cepheids they contained.

The scale model of the Universe thus became a real map. The Cepheid variables had become flickering yardsticks in the hands of the astronomers.

In 1918 Harlow Shapley started calculating the distance of the various globular clusters from the RR Lyrae variables they contained, using Miss Leavitt's period-luminosity curve. His figures turned out to be a little too large and were corrected downward during the next decade, but the new picture of the Galaxy which grew out of his measurements has survived.

The globular clusters are distributed spherically above and below the median plane of the Galaxy. The centre of this sphere of globular clusters is in the plane of the Galactic lens, but at a spot some tens of thousands of light-years from us in the direction of the constellation Sagittarius.

That explained why most of the globular clusters were to be found in that direction.

It seemed to Shapley a natural assumption that the globular clusters centred about the centre of the Galaxy and later evidence from other directions bore him out. So here we are, not in the centre of the Galaxy at all, but well out to one side.

We are still in the median plane of the Galaxy, for the Milky Way does split the heavens in half. But how account for the fact that the Milky Way is equally bright throughout if we are not, in fact, centred? The answer is that the median plane on the outskirts of the Galaxy (where we happen to be) is loaded with dust clouds. These happen to lie between ourselves and the Galactic centre, obscuring it completely.

The result is that, with or without optical telescopes, we can only see our portion of the Galactic outskirts. We *are* in the centre of that part of the Galaxy which we can see optically, and that part is not too far off in size from Kapteyn's estimate. Kapteyn's error (which was at the time excusable) was in assuming that what we could see was all the Galaxy there was. It wasn't.

The final model of the Galaxy, now thought to be correct, is that of a lens-shape that is 100,000 light-years across and 20,000 light-years thick at the centre. The thickness falls off as the edge is approached and in the position of the Sun (30,000 light-years from the centre and two-thirds of the way towards the extreme edge of the Galaxy) is only 3,000 light-years thick.

Even before the Galactic measurements had been finally determined, the Cepheids in the Magellanic Clouds had been used to determine their distance. Those turned out to be rather more than 100,000 light-years distant. (Our best modern figures are 150,000 light-years for the Large and 170,000 light-years for the Small Magellanic Cloud.) They are close enough to the body of our Galaxy and small enough in comparison, to be fairly considered 'satellite Galaxies' of ours.

From the rate at which our Sun and neighbouring stars are travelling in their 200,000,000-year circuit of the Galactic centre, it is possible to calculate the mass of the Galactic centre (which contains most of the stars in the Galaxy) and it turns out to be something like 90,000,000,000 times that of our Sun. If we consider our Sun to be an average star in mass, then we can fairly estimate the Galaxy as a whole to contain 100,000,000,000 stars. In compariosn, the two Magellanic Clouds together contain a total of about 6,000, 000,000.

The question in the 1920s was whether anything existed in the Universe outside our Galaxy and its satellites. Suspicion rested on certain dim, foggy structures, of which a cloudy patch in the constellation Andromeda was the most spectacular. (It was about half the size of the full moon to the naked eye and was called the Andromeda nebula — 'nebula' coming from a Greek word for cloud.)

There were some nebulae which were known to be parts of our Galaxy because they contained hot (and not too distant) stars that were the cause of their luminosity. The Orion nebula is an example. The Andromeda nebula, however, contained no such stars that anyone could see and seemed to be self-luminous. Could it be, then, a patch of haze that could be broken up into many far, far distant stars (with the proper magnification), as could the Milky Way and the Magellanic Clouds? Since the same telescopes that resolved the Milky Way and the Magellanic Clouds did not manage to do the same for the Andromeda nebula, could it be that the Andromeda nebula was far more distant?

The answer came in 1924, when Edwin Powell Hubble turned the new 100-inch telescope at Mount Wilson on the Amdromeda nebula and took photographs that showed the outskirts of the nebula resolved into stars. Furthermore, he found Cepheids among the newly revealed stars and used them to determine the distance. The Andromeda nebula turned out to be 750,000 light-years distant and this is the value found in all the astronomy books published over the next thirty years.

Allowing for the distance, the Andromeda nebula was obviously an object of galactic size, so it is now called the Andromeda galaxy. Hubble established the fact that many other nebulae of the Andromeda type were galaxies, even more distant than the Andromeda galaxy (which is a near neighbour of ours, in fact). The size of the Universe sprang instantly from a diameter in the hundreds of thousands of light-years to one in the hundreds of millions.

However, there were a few disturbing facts which lingered. For one thing, the other galaxies all seemed to be considerably smaller than our own. Why should our own Galaxy be the one outsize member of a large group?

For another, the Andromeda galaxy had a halo of globular clusters just as our own Galaxy did. Their clusters, however, were considerably smaller and dimmer than ours. Why?

Thirdly, allowing for the distance of the galaxies and the

speed at which the Universe was known to be expanding it could only have been two billion years ago that all the galaxies were squashed together at some central starting point. The trouble with that was that the geologists swore up and down that the earth itself was considerably older than two billion years. How could the earth be older than the Universe?

The beginning of an answer came in 1942, when Walter Baade took another look at the Andromeda galaxy with the 100-inch telescope. Until then only the outer fringes of the galaxy had been broken up into stars; the central portions had remained a featureless fog. But now Baade had an unusual break. It was wartime and Los Angeles was blacked out. That removed the dim background of distant city light and improved 'seeing'.

For the first time, photographs were taken that resolved the inner portions of the Andromeda galaxy. Baade could study the very brightest stars of the interior.

It turned out that there were striking differences between the brightest stars of the inner regions and those of the outskirts. The brightest stars in the interior were reddish while those of the outskirts were bluish. That alone accounted for the greater ease with which photographic plates picked up the outer stars, since blue more quickly affects the plates than red does (unless special plates are used). Add to this, the fact that the brightest (bluish) stars on the outskirts were up to a hundred times as bright as the brightest (reddish) stars in the interior.

To Baade, it seemed there were two sets of stars in the Andromeda galaxy with different structures and history. He called the stars of the outskirts Population I, and those of the interior Population II.

Population II is the dominant star-group of the Universe, making up perhaps 98 per cent of the total. They are, by and large, old, moderate-sized stars fairly uniform in characteristics and moving about in dust-free surroundings.

Population I stars are found only in the dust-choked spiral arms of those galaxies that have spiral arms. On the whole, they are far more scattered in age and structure than the Population II stars, including very young, hot, and luminous stars. (Perhaps Population I stars sweep up the dust through which they pass gradually growing more massive, hotter, and brighter — and shortening their lives, as humans do, by overeating.)

Our own Sun, by the way, happens to be occupying a spiral arm so that the familiar stars of our sky belong to Population I. Our own Sun, fortunately, is an old, quiet, settled star not typical of that turbulent group.

Once the 200-inch telescope was set up on Mount Palomar, Baade continued his investigations of the two populations. There were Cepheid variables in both populations and this brought up an interesting point.

The Cepheids of the Magellanic Clouds (which have no spiral arms) belong to Population II. So do the RR Lyrae variables in the globular clusters. So do the Cepheids of the moderately distant non-globular clusters for which the actual distances were first worked out statistically. In other words all the work done on the size of the Galaxy and the distance of the Magellanic Clouds, as well as on the original establishment of the Cepheid yardstick, were done on Population II Cepheids. So far, so good.

But what about the distance of the outer galaxies? The only stars that could be made out in the outer galaxies such as Andromeda by Hubble and his successors were the extralarge giants of the spiral arms. Those extra-large giants were Population I, and the Cepheids among them were Population I Cepheids. Since Population I is so different from Population II, could one be certain that the Population I

Cepheids would fit into a period-luminosity curve which had been worked out from Population II Cepheids only?

Baade began a painstaking comparison of the Population II Cepheids in the globular clusters with the Population I Cepheids in our own neighbourhood and in 1952 announced that the latter did *not* fit the Leavitt period-luminosity curve. For any particular period, a Population I Cepheid was between four and five times as luminous as a Population II Cepheid would be. A new period-luminosity curve was drawn for the Population I Cepheids.

Well, then, if the Population I Cepheids of the Andromeda spiral arms were each considerably more than four times brighter than had been thought, then to be as bright as they seemed (the apparent brightness stayed the same of course) they had to be considerably more than twice as far away as had been thought. The flickering yardstick of the Cepheids, which the astronomers had been using to measure the distance of the outer galaxies suddenly turned out to be roughly three times as long as they had thought.

All the nearer galaxies, which had been measured by that yardstick were suddenly pushed a triple distance out into space. The further galaxies whose distances had been estimated by procedures based on the 'known' distances of the nearer galaxies, all retreated likewise.

The Universe had again increased in size, and the 200-inch telescope was penetrating, not somewhat less than a billion light-years into space, but a full two billion light-years.

This solved the puzzles of the galaxies. If all the galaxies were about three times the distance that had been thought, they must be larger (in actuality) than had been thought. With all the galaxies suddenly grown up, our own Galaxy is reduced to run-of-the-mill size and is no longer the one outsize member of the family. In fact, the Andromeda galaxy is at least twice the size of ours in terms of numbers of stars contained.

Secondly, the globular clusters around Andromeda, being actually much further away than had been thought, must be more luminous in actual fact than had been thought. Once the greater distance had been allowed for, the globular clusters of Andromeda worked out to be quite comparable to our own in brightness.

Finally, with the galaxies much further spread apart, but with their actual speeds of recession unaffected by the change (the measurement of speed of recession does not depend on the distance of the object being tested) it would take a much longer period for the Universe to have reached its present state from an original compressed hunk of matter. This meant the age of the Universe had to be, at minimum, five or even six billion years. With this figure, geologists were content. They no longer had to consider the Earth to be older than the Universe.

Which was a great relief.

CHAPTER FIFTEEN

THE SIGHT OF HOME

Now man is struggling towards the Moon[1] but someday, we hope, he will be bouncing among the far stars. Can we imagine that the time may come when some homesick astronaut will lift his eyes to the strange skies of planets of distant suns in order to locate the tiny speck that is 'Sol'? — home, sweet home, across the frigid vastness of space.

A touching picture, but what occurs to me is: How far away can said astronaut be and still make out the sight of home? For that matter, we can make it general and ask: How far away can the inhabitant of any stellar system be and still make out the sight of the star in whose planetary system he was born?

This, of course, depends on how bright the particular star is. I say *is* and not *seems*. From where we sit here on Earth's surface, we see stars of all gradations of brightness. That brightness is partly due to the star's particular luminosity, but is also partly due to the distance it happens to be from us. A star not particularly bright, as stars go, might seem a brilliant specimen to us because it is relatively close; while a star much brighter, but also much more distant, might seem trivial in comparison.

Consider the two stars Alpha Centauri and Capella, for instance. Both are about equally bright in appearance, with magnitudes of about 0·1 and 0·2 respectively. (Remember that the lower the magnitude, the brighter the star, and that each unit decrease of magnitude is equal to a multiplication of 2·52 in brightness.)

However, the two stars are not the same distance from us. Alpha Centauri is the closest of all stars and is only 1·3 parsecs from us. (I am giving all distances in parsecs in this chapter for a reason I will shortly explain. To guide you, a parsec is equal to 3·26 light-years, or to 19,150,000,000,000 miles.) Capella, on the other hand, is 14 parsecs from us, or over 10 times the distance of Alpha Centauri.

Since the intensity of light decreases as the square of the distance, the light of Capella has had a chance to decrease by 10×10 or 100 times more than has the light of Alpha Centauri. Since Capella ends by appearing as bright as Alpha Centauri, it must in reality be 100 times as bright.

If we know a star's distance, we can correct for it. We can calculate what its brightness would be if it were located at some standard distance from us. The distance actually used by astronomers as standard in this connection is 10 parsecs (which is why I am giving all distances as parsecs in this chapter).

Thus the *apparent magnitude* (the actual brightness of a star as we see it) of Alpha Centauri is 0·1 and of Capella is 0·2. The *absolute magnitude*, however (the brightness as it would appear if a star were exactly 10 parsecs away) is 4·8 for Alpha Centauri and −0·6 for Capella.

The Sun, by the way, is just about as bright as Alpha Centauri. Its absolute magnitude is 4·86. Both are average, run-of-the-mill stars.

It is possible to relate absolute magnitude, apparent magnitude, and distance by means of Equation 16.

$$M = m + 5 - 5 \log D \qquad \text{(Equation 16)}$$

where M is the absolute magnitude of a star, m is the apparent magnitude and D is the distance in parsecs. At the standard distance of 10 parsecs the value of D is 10 and log 10 equals 1. The equation becomes $M = m+5-5$, or $M = m$. The equation at least checks by telling us that at the standard distance of 10 parsecs the apparent magnitude is equal to the absolute magnitude.

But let's use the equation for something more significant. Our astronaut is on a planet of another star and he wants to point out the Sun to the local gentry. He wants to do so with pride so he would like to have it a first-magnitude star.

The equation will tell us how far away we can be in order that this might be possible. The absolute magnitude of the Sun (M) is 4·86. That can't be changed. We want the apparent magnitude to be 1, so we substitute that for m. We now calculate for D which turns out to be equal to 1·7 parsecs.

Only Alpha Centauri is within 1·7 parsecs of the Sun. This means from a planet in the Alpha Centauri system only can the Sun be seen as a first-magnitude star, and from no other planetary system in the Universe. Sirius, for instance, is very close to us (less than 3 parsecs away, close enough to be incomparably the brightest star in the sky though only ½ as bright as Capella in actuality) and yet even from the Sirian system, the Sun would be seen as only a second-magnitude star.

Well, then, his pride chastened, but homesick nevertheless, our astronaut might abandon first-magnitude pretensions and be willing to settle for any glimpse, however faint, of home.

Since a star of apparent magnitude 6·5 can just barely be made out by a pair of excellent eyes under ideal seeing condition, let's make m equal to 6·5 instead of to 1 and calculate a new value for D. Now it comes out as equal to

20 parsecs. The Sun is down to the very limit of naked-eye visibility at a distance of 20 parsecs.

Of course, it is visible for this distance in all directions (assuming that it is not obscured by dust clouds or anything like that) so that it can be seen by naked eye anywhere in a sphere of which the Sun is the centre and which has a radius of 20 parsecs. The volume of such a sphere is about 32,000 cubic parsecs.

This sounds like a lot but in the neighbourhood of our Sun, the density of stars (or multiple stars) is about $4\frac{1}{2}$ per 100 cubic parsecs. Within the visibility sphere of the Sun there are therefore about 1,450 stars or multiple-star systems. Since the Galaxy contains about a hundred billion stars, the number of stellar systems from which we can be seen at all, by naked eye, represents an insignificant percentage of those in the Galaxy.

Or put it another way. The Galaxy is about 30,000 parsecs across the full width of its lens-shape. The range of visibility of the Sun is only $1/800$ of this.

Obviously, if we are going to go flitting here and there in the Galaxy, we can just take it for granted that when we lift our tear-filled, homesick eyes to the alien heavens, a sight of home is what we will not get.

Of course, let's not be too sorry for ourselves. There are stars far less luminous than the Sun and therefore far less extensively visible.

The least luminous star known is one which is listed in the books as 'Companion of BD + 4°4048' which, for obvious reasons I suggest we call (for purposes of this chapter only) Joe. Now Joe has an absolute magnitude of 19·2. It is only two millionths as bright as the Sun and although it is only about 6 parsecs from us, it is barely visible in a good large telescope.

Using the equation, it turns out that at a distance of 0·03 parsecs, Joe is just barely visible to the naked eye. This means that if Joe were put in the place of the Sun, it would disappear from naked-eye sight at a distance just six times as great as that of the planet Pluto.

It is unlikely that anywhere in the Galaxy there exist two stars this close together, unless, of course, they form part of a multiple-star system. (And Joe *is* part of a multiple-star system, one which includes the star BD+4°4048, of which it is the 'Companion'.)

It follows that the existence of a star like Joe would be a complete secret to any race of beings not possessing telescopes and not living on a planet that actually revolves about Joe or about its companion. No man from Joe could ever get a naked-eye sight of home from any planet outside his own multiple system; from any planet at all.

On the other hand, consider stars brighter than the Sun. Sirius, with an absolute magnitude of 1·36 can be made out at a distance of 100 parsecs, while Capella with an absolute magnitude of −0·6 could be seen as far off as 260 parsecs. Sirius could be seen through a volume of space 600 times and Capella through a volume over 2,000 times as great as that through which the Sun can be seen.

Nor is Capella the most luminous star by any means. Of all the stars visible to the naked eye, Rigel is about the most luminous. It has an absolute magnitude of −5·8, which makes it over 20,000 times as luminous as the Sun and rather more than 100 times as luminous as even bright Capella.

Rigel can be seen by the naked eye for a distance of 2,900 parsecs in any direction, which means over a range of $\frac{1}{5}$ the width of the Galaxy. This is rather respectable.

It means that over a large section of the Galaxy we might at least count on identifying our Sun by its spectacular neighbour. We could say to the local Rotarians, 'Oh, well you can't see our Sun from here, but it's pretty close to Rigel, that star over there, the one you call Bjfxlpt.'

But the record for steady day-in and day-out luminosity is not held by any member of our own Galaxy. There is a star called S Doradus in the Large Magellanic Cloud (which is a kind of satellite galaxy of our own, about 50,000 parsecs away) and S Doradus has an absolute magnitude of −9. It can be seen by naked eye for a distance of 12,500 parsecs. It could be made out all across its own small galaxy and across nearly the full length of our own large one, if it were in our Galaxy.

Of course, no normal star can compete in brightness with a star that explodes. Exploding stars fall into two classes. First there are ordinary novae, which every million years or so blow off a per cent or so of their mass and grow several thousand times brighter (temporarily) when they do so. In between blowoffs they lead fairly normal lives as ordinary stars. Such novae may reach absolute magnitudes of −9, which makes them only as bright as S Doradus is all the time, but then S Doradus is a most unusual star. Certainly the novae are a million times as luminous as are average stars like our Sun.

But then there are supernovae. These are stars that go completely to smash in one big explosion, releasing as much energy in a second as the Sun does in sixty years. Most of their mass is blown off and what is left is converted to a white dwarf.[2] The upper limit of their absolute magnitude reaches anywhere between −14 and −17 so that a large supernova can be 1,500 times as luminous as even S Doradus.

If we imagined a good supernova reaching an absolute magnitude of −17, it could be seen by naked eye, at peak brightness, for a distance of 500,000 parsecs. In other words, such a supernova flaring up anywhere in our Galaxy could be seen by naked eye anywhere else in our Galaxy (except where obscured by interstellar dust). It could even be seen in our satellite galaxies, the Large and Small Magellanic Clouds.

However, the distance between our Galaxy and the nearest full-sized neighbour, the Andromeda galaxy, is about 700,000 parsecs. It follows that supernova in other galaxies cannot be seen by naked eye. Any supernova that is visible by naked eye must be located in our own Galaxy, or, at most, in the Magellanic Clouds.

Now astronomers have studied novae which have flared up in our Galaxy. For instance there was a nova in the constellation Hercules in 1934 that rose from telescopic obscurity to the 2nd magnitude (say as bright as the North Star) in a matter of days and stayed near that brightness for three months. In 1942, a nova reached first magnitude (as bright as Arcturus) for a month.

But novae themselves aren't unusual. An average of 20 flare out per year per galaxy.

Supernovae are different breeds altogether and astronomers would love to get data on them. Unfortunately, they are quite rare. It is estimated that about 3 supernovae appear per galaxy per millennium; that is, one supernova for every 7,000 ordinary novae.

Naturally, a supernova can be best studied if it appears in our own Galaxy, and astronomers are waiting for one to appear.

Actually, there is a chance that our Galaxy has had its expected 3 supernovae in the course of the last thousand years. At least there were three very bright novae which have been sighted by naked eye in that interval.

The first of these was sighted in A.D. 1054 by Chinese and Japanese astronomers. From the position in the constellation Taurus, recorded by these Orientals, modern astronomers had a pretty good notion as to where to look for any remnants of the novae. In 1844 the English astronomer, William Parsons, located an odd object in the appropriate place. It was a tiny star barely visible in a good telescope (which eventually turned out to be a white dwarf[3]). Surrounding it was an irregular mass of glowing gas. Because the gas was irregular, with clawlike projections, the object was named the Crab Nebula.

Continued observation over decades showed the gas was expanding. Spectroscopic data revealed the true rate of expansion and that combined with the apparent rate revealed the distance of the Crab Nebula to be about 1,600 parsecs. Assuming that the gas had been exploded outward at some time in the past, it was possible to calculate backward to see when that explosion had taken place (from the present position and rate of expansion of the gas). It turns out the explosion took place about 900 years ago. There seems no doubt that the Crab Nebula is what remains of the nova of 1054.

For the nova to be brighter than Venus it must have had a peak apparent magnitude of −5. Substituting that for m in the equation and 1,600 for D, the value of M, the absolute magnitude, works out to be just about −16. From this and from the white-dwarf remnant and the gassy explosion, there can be no doubt that the nova of 1054 was a true supernova and one which took place within our Galaxy.

In 1572 a new star appeared in the constellation Cassiopeia. It also outshone Venus and was visible by day. This time it was observed by Europeans. In fact, the last and most famous of all naked-eye astronomers, Tycho Brahe, observed it as a young man and wrote a book about it entitled *De Nova Stella* ('Concerning the New Star') and it is from that title that the word 'nova' for new stars comes.

In 1604 still another new star appeared, this time in the

constellation Serpens. It was not quite as bright as the nova of 1572 and perhaps only grew as bright as Mars at its brightest (say an apparent magnitude of −2·5). It was observed by another great astronomer, Johann Kepler, who had been Tycho's assistant in the latter's final years.

Now the question is, were the novae of 1572 and 1604 supernovae? Unlike the case of the nova of 1054, no white dwarf, no nebulosity, no anything has been located in the spots reported by Tycho and Kepler. The direct evidence of supernova-hood is missing. Perhaps they were only ordinary novae.

Well, if they were ordinary novae with absolute magnitudes of only −9, then the nova of 1572 must have been about 60 parsecs distant, no more, if it was to surpass Venus in brightness. The nova of 1604 would be 200 parsecs distance. Stars that close could scarcely fail to be seen with modern telescopes, even if they were dim, it seems to me. (Of course, if they ended up as dim as 'Joe' they might not be seen, but that level of dimness is most unlikely.[4])

Most astronomers seem satisfied that the novae of 1572 and 1604 were supernovae in our own Galaxy, and this brings up an irony of astronomical history. Two supernovae appeared in the space of a single generation, the generation just before the invention of the telescope, and not one supernova has appeared in our Galaxy in the nine generations since.

Even a small telescope could have plotted the position of the supernovae more exactly and made it somewhat more likely that the remnant could now be located. Then if the supernovae had appeared after the invention of the spectroscope, things would have been rosier still for happy little astronomers.

As it is, supernovae *have* been observed since Kepler's time, about 50 altogether, but only in other galaxies, so that the apparent brightness was so low that little detail could be made out in the spectra.

The brightest and closest supernova to have appeared since 1604 showed up in 1885 in the Andromeda galaxy, our neighbour. It reached an apparent magnitude of 7. (It was not quite visible, you will note, to the naked eye. As I said before, only supernovae in our own Galaxy or in the Magellanic Clouds are visible to the naked eye.) Since the Andromeda galaxy lies at a distance of 700,000 parsecs, the absolute magnitude of the supernova comes out to just a bit brighter than -17. It was about a tenth as bright as the entire galaxy that contained it. Also, since the Andromeda galaxy is considerably larger than our own, you might say that this one supernova approached the brightness of all stars of the Milky Way put together — for a while anyway.

(In fact, it was the extraordinary brightness of this star that eventually made astronomers realise that there were nova that were thousands of times brighter than run-of-the-mill nova, and thus the concept of the supernova arose.)

Well now, telescopes and spectroscopes were trained on the supernova of 1885 so that it was better studied than were the much closer ones of 1572 and 1604, but astronomers still weren't living right. Photography had not yet been applied to spectroscopy. If the supernova of 1885 had held on for 20 years more, or if it had been located 20 light-years further from Earth (so that the light would have taken 20 years longer to reach us) its spectrum could have been recorded photographically and studied in detail.

Well, astronomers can only wait! It's even money that sometime during the next century, there will be a supernova blowing its top either in our Galaxy or in the Andromeda galaxy and this time cameras (and heaven knows what else — radio telescopes, too) will be waiting. Provided, of course, the next supernova is not old Sol — the chance of which is, however, virtually nil from what little we know of super-novae.

Still that does bring up a rather grisly situation — the doom of the Earth by nova-formation on the part of the Sun. The Earth would be puffed into gas within minutes of an explosion of the Sun.

Yet is it only the Sun that need concern us? What if a neighbouring star exploded?

For instance, suppose it was Alpha Centauri that decided to blow up. If Alpha Centauri became an ordinary nova and reached an absolute magnitude of -9, then its apparent magnitude would be $-13\cdot5$. It would be two and a half times as bright as the full Moon and a fine spectacle for that portion of Earth's population that lived farther south than Florida and Egypt. (It would be a new tourist attraction and countries like Argentina, the Union of South Africa, and Australia would clean up for a few months.)

Or suppose Alpha Centauri went supernova and reached an absolute magnitude of -17. (It is impossible for it to do so according to current theories, but let's suppose it anyway.) Its apparent magnitude would then be $-21\cdot5$ which would make it 4,000 times as bright as the full Moon and actually $1/_{160}$ as bright as the Sun.

Under such conditions there would be no night for any region of the Earth that had Alpha Centauri in its night sky. You could read newspapers and you would cast a shadow. With Alpha Centauri in the day sky, it would still be a clearly visible and blindingly brilliant point of light and, in the absence of clouds, you would cast a double shadow. In fact, for a couple of months, Earth would be truly a planet of a double sun.

The total amount of energy reaching Earth would (temporarily) be increased by as much as 0·6 per cent. This might have a significant effect on the weather. A large part of the Alpha Centauri radiation would be high-energy and that ought to play hob with the upper atmosphere. In short, although Alpha Centauri as supernova might not endanger life on Earth, it would certainly make things hot for us for a while.

CHAPTER SIXTEEN

THE BLACK OF NIGHT

I suppose many of you are familiar with the comic strip 'Peanuts'. My daughter Robyn (now in the fourth grade)[1] is very fond of it, as I am myself.

She came to me one day, delighted with a particular sequence in which one of the little characters in 'Peanuts' asks his bad-tempered older sister, 'Why is the sky blue?' and she snaps back, 'Because it isn't green!'

When Robyn was all through laughing, I thought I would seize the occasion to manoeuvre the conversation in the direction of a deep and subtle scientific discussion (entirely for Robyn's own good, you understand). So I said, 'Well, tell me, Robyn, why is the night sky black?'

And she answered at once (I suppose I ought to have foreseen it), 'Because it isn't purple!'

Fortunately, nothing like this can ever seriously frustrate me. If Robyn won't co-operate, I can always turn, with a snarl, on the Helpless Reader. I will discuss the blackness of the night sky with *you*!

The story of the black of night begins with a German physician and astronomer, Heinrich Wilhelm Matthias Olbers, born in 1758. He practised astronomy as a hobby, and in midlife suffered a peculiar disappointment. It came about in this fashion ...

Towards the end of the eighteenth century, astronomers began to suspect, quite strongly, that some sort of planet must exist between the orbits of Mars and Jupiter. A team of German astronomers, of whom Olbers was one of the most important, set themselves up with the intention of dividing the ecliptic among themselves and each searching his own portion, meticulously, for the planet.

Olbers and his friends were so systematic and thorough that by rights they should have discovered the planet and received the credit of it. But life is funny (to coin a phrase). While they were still arranging the details, Giuseppe Piazzi, an Italian astronomer who wasn't looking for planets at all, discovered, on the night of January 1st, 1801, a point of light which had shifted its position against the background of stars. He followed it for a period of time and found it was continuing to move steadily. It moved less rapidly than Mars and more rapidly than Jupiter, so it was very likely a planet in an intermediate orbit. He reported it as such so that it was the casual Piazzi and not the thorough Olbers who got the nod in the history books.

Olbers didn't lose out altogether, however. It seems that after a period of time, Piazzi fell sick and was unable to continue his observations. By the time he got back to the telescope the planet was too close to the Sun to be observable.

Piazzi didn't have enough observations to calculate an orbit and this was bad. It would take months for the slow-moving planet to get to the other side of the Sun and into observable position, and without a calculated orbit it might easily take years to rediscover it.

Fortunately, a young German mathematician, Karl Friedrich Gauss, was just blazing his way upward into the mathematical firmament. He had worked out something called the 'method of least squares', which made it possible to calculate a reasonably good orbit from no more than three good observations of a planetary position.

Gauss calculated the orbit of Piazzi's new planet, and when it was in observable range once more there was Olbers and his telescope watching the place where Gauss's calculations said it would be. Gauss was right and, on January 1st, 1802, Olbers found it.

To be sure, the new planet (named 'Ceres') was a peculiar one, for it turned out to be less than 500 miles in diameter. It was far smaller than any other known planet and smaller than at least six of the satellites known at that time.

Could Ceres be all that existed between Mars and Jupiter? The German astronomers continued looking (it would be a shame to waste all that preparation) and sure enough, three more planets between Mars and Jupiter were soon discovered. Two of them, Pallas and Vesta, were discovered by Olbers. (In later years many more were discovered.)

But, of course, the big payoff isn't for second place. All Olbers got out of it was the name of a planetoid. The thousandth planetoid between Mars and Jupiter was named 'Piazzia', the thousand and first 'Gaussia', and the thousand and second (hold your breath, now) 'Olberia'.

Nor was Olbers much luckier in his other observations. He specialised in comets and discovered five of them, but practically anyone can do that. There is a comet called 'Olbers' Comet' in consequence, but that is a minor distinction.

Shall we now dismiss Olbers? By no means.

It is hard to tell just what will win you a place in the annals of science. Sometimes it is a piece of interesting reverie that does it. In 1826 Olbers indulged himself in an idle speculation concerning the black of night and dredged out of it an apparently ridiculous conclusion.

Yet that speculation became 'Olbers' paradox', which has

come to have profound significance a century afterwards. In fact, we can begin with Olbers' paradox and end with the conclusion that the only reason life exists anywhere in the universe is that the distant galaxies are receding from us.

What possible effect can the distant galaxies have on us? Be patient now and we'll work it out.

In ancient times, if any astronomer had been asked why the night sky was black, he would have answered — quite reasonably — that it was because the light of the Sun was absent. If one had then gone on to question him why the stars did not take the place of the Sun, he would have answered — again reasonably — that the stars were limited in number and individually dim. In fact, all the stars we can see would, if lumped together, be only a half-billionth as bright as the Sun. Their influence on the blackness of the night sky is therefore insignificant.

By the nineteenth century, however, this last argument had lost its force. The number of stars was tremendous. Large telescopes revealed them by the countless millions.

Of course, one might argue that those countless millions of stars were of no importance for they were not visible to the naked eye and therefore did not contribute to the light in the night sky. This, too, is a useless argument. The stars of the Milky Way are, individually, too faint to be made out, but *en masse* they make a dimly luminous belt about the sky. The Andromeda galaxy is much farther away than the stars of the Milky Way and the individual stars that make it up are not individually visible except (just barely) in a very large telescope. Yet, *en masse*, the Andromeda galaxy is faintly visible to the naked eye. (It is, in fact, the farthest object visible to the unaided eye; so if anyone ever asks you how far you can see; tell him 2,000,000 light-years.)

In short, distant stars — no matter how distant and no matter how dim, individually — must contribute to the light of the night sky, and this contribution can even become detectable without the aid of instruments if these dim distant stars exist in sufficient density.

Olbers, who didn't know about the Andromeda galaxy, but did know about the Milky Way, therefore set about asking himself how much light ought to be expected from the distant stars altogether. He began by making several assumptions:

1. That the universe is infinite in extent.

2. That the stars are infinite in number and evenly spread throughout the universe.

3. That the stars are of uniform average brightness through all of space.

Now let's imagine space divided up into shells (like those of an onion) centring about us, comparatively thin shells compared with the vastness of space, but large enough to contain stars within them.

Remember that the amount of light that reaches us from individual stars of equal luminosity varies inversely as the square of the distance from us. In other words, if Star A and Star B are equally bright but Star A is three times as far as Star B, Star A delivers only $1/9$ the light. If Star A were five times as far as Star B, Star A would deliver $1/25$ the light, and so on.

This holds for our shells. The average star in a shell 2,000 light-years from us would be only $\frac{1}{4}$ as bright in appearance as the average star in a shell only 1,000 light-years from us. (Assumption 3 tells us, of course, that the intrinsic brightness of the average star in both shells is the same, so that distance is the only factor we need to consider.) Again, the average star in a shell 3,000 light-years from us would be only $1/9$ as bright in appearance as the average star in the 1,000-light-year shell, and so on.

But as you work your way outward, each succeeding shell is more voluminous than the one before. Since each shell is thin enough to be considered, without appreciable error, to be the surface of the sphere made up of all the shells within, we can see that the volume of the shells increases as the surface of the spheres would — that is, as the square of the radius. The 2,000-light-year shell would have four times the volume of the 1,000-light-year shell. The 3,000-light-year shell would have nine times the volume of the 1,000-light-year shell, and so on.

If we consider the stars to be evenly distributed through space (Assumption 2), then the number of stars in any given shell is proportional to the volume of the shell. If the 2,000-light-year shell is four times as voluminous as the 1,000-light-year shell, it contains four times as many stars. If the 3,000-light-year shell is nine times as voluminous as the 1,000-light-year shell, it contains nine times as many stars, and so on.

Well, then, if the 2,000-light-year shell contains four times as many stars as the 1,000-light-year shell, and if each star in the former is $\frac{1}{4}$ as bright (on the average) as each star of the latter, then the total light delivered by the 2,000-light-year shell is 4 times $\frac{1}{4}$ that of the 1,000-light-year shell. In other words, the 2,000-light-year shell delivers just as much total light as the 1,000-light-year shell. The total brightness of the 3,000-light-year shell is 9 times $1/9$ that of the 1,000-light-year shell, and the brightness of the two shells is equal again.

In summary, if we divide the universe into successive shells, each shell delivers as much light, *in toto*, as do any of the others. And if the universe is infinite in extent (Assumption 1) and therefore consists of an infinite number of shells, the stars of the universe, however dim they may be individually, ought to deliver an infinite amount of light to the Earth.

The one catch, of course, is that the nearer stars may block the light of the more distant stars.

To take this into account, let's look at the problem in another way. In no matter which direction one looks, the eye will eventually encounter a star, if it is true they are infinite in number and evenly distributed in space (Assumption 2). The star may be individually invisible, but it will contribute its bit of light and will be immediately adjoined in all directions by other bits of light.

The night sky would then not be black at all but would be an absolutely solid smear of starlight. So would the day sky be an absolutely solid smear of starlight, with the Sun itself invisible against the luminous background.

Such a sky would be roughly as bright as 150,000 suns like ours, and do you question that under those conditions life on Earth would be impossible?

However, the sky is *not* as bright as 150,000 suns. The night sky *is* black. Somewhere in the Olbers' paradox there is some mitigating circumstance or some logical error.

Olbers himself thought he found it. He suggested that space was not truly transparent; that it contained clouds of dust and gas which absorbed most of the starlight, allowing only an insignificant fraction to reach the Earth.

That sounds good, but it is no good at all. There are indeed dust clouds in space but if they absorbed all the starlight that fell upon them (by the reasoning of Olbers' paradox) then their temperature would go up until they grew hot enough to be luminous. They would, eventually, emit as much light as they absorb and the Earth sky would still be star-bright over all its extent.

But if the logic of an argument is faultless and the conclusion is still wrong, we must investigate the assumptions. What about Assumption 2, for instance? Are the stars indeed infinite in number and evenly spread throughout the universe?

Even in Olbers' time there seemed reason to believe this assumption to be false. The German-English astronomer William Herschel made counts of stars of different brightness. He assumed that, *on the average*, the dimmer stars were more distant than the bright ones (which follows from Assumption 3) and found that the density of the stars in space fell off with distance.

From the rate of decrease in density in different directions, Herschel decided that the stars made up a lens-shaped figure. The long diameter, he decided, was 150 times the distance from the Sun to Arcturus (or 6,000 light-years, we would now say), and the whole conglomeration would consist of 100,000,000 stars.

This seemed to dispose of Olbers' paradox. If the lens-shaped conglomerate (now called the Galaxy) truly contained all the stars in existence, then Assumption 2 breaks down. Even if we imagined space to be infinite in extent outside the Galaxy (Assumption 1), it would contain no stars and would contribute no illumination. Consequently, there would be only a finite number of star-containing shells and only a finite (and not very large) amount of illumination would be received on Earth. That would be why the night sky is black.

The estimated size of the Galaxy has been increased since Herschel's day. It is now believed to be 100,000 light-years in diameter, not 6,000; and to contain 150,000,000,000 stars, not 100,000,000. This change, however, is not crucial, it still leaves the night sky black.

In the twentieth century Olbers' paradox came back to

life, for it came to be appreciated that there were indeed stars outside the Galaxy.

The foggy patch in Andromeda had been felt throughout the nineteenth century to be a luminous mist that formed part of our own Galaxy. However, other such patches of mist (the Orion Nebula, for instance) contained stars that lit up the mist. The Andromeda patch, on the other hand, seemed to contain no stars but to glow of itself.

Some astronomers began to suspect the truth, but it wasn't definitely established until 1924, when the American astronomer Edwin Powell Hubble turned the 100-inch telescope on the glowing mist and was able to make out separate stars in its outskirts. These stars were individually so dim that it became clear at once that the patch must be hundreds of thousands of light-years away from us and far outside the Galaxy. Furthermore, to be seen, as it was, at that distance, it must rival in size our entire Galaxy and be another galaxy in its own right.

And so it is. It is now believed to be over 2,000,000 light-years from us and to contain at least 200,000,000,000 stars. Still other galaxies were discovered at vastly greater distances. Indeed, we now suspect that within the observable universe there may be as many as 100,000,000,000 galaxies, and the distance of some of them has been estimated as high as 6,000,000,000 light-years.

Let us take Olbers' three assumptions then and substitute the word 'galaxies' for 'stars' and see how they sound.

Assumption 1, that the universe is infinite, sounds good. At least there is no sign of an end even out to distances of billions of light-years.

Assumption 2, that *galaxies* (not stars) are infinite in number and evenly spread throughout the universe, sounds good, too. At least they are evenly distributed for as far out as we can see, and we can see pretty far.

Assumption 3, that *galaxies* (not stars) are of uniform average brightness throughout space, is harder to handle. However, we have no reason to suspect that distant galaxies are consistently larger or smaller than nearby ones, and if the galaxies come to some uniform average size and star content, then it certainly seems reasonable to suppose they are uniformly bright as well.

Well, then, why is the night sky black? We're back to that.

Let's try another tack. Astronomers can determine whether a distant luminous object is approaching us or receding from us by studying its spectrum (that is, its light as spread out in a rainbow of wavelengths from short-wavelength violet to long-wavelength red).

The spectrum is crossed by dark lines which are in a fixed position if the object is motionless with respect to us. If the object is approaching us, the lines shift towards the violet. If the object is receding from us, the lines shift towards the red. From the size of the shift astronomers can determine the velocity of approach or recession.

In the 1910s and 1920s the spectra of some galaxies (or bodies later understood to be galaxies) were studied, and except for one or two of the very nearest, all are receding from us. In fact, it soon became apparent that the farther galaxies are receding more rapidly than the nearer ones. Hubble was able to formulate what is now called 'Hubble's Law' in 1929. This states that the velocity of recession of a galaxy is proportional to its distance from us. If Galaxy A is twice as far as Galaxy B, it is receding at twice the velocity. The farthest observed galaxy, 6,000,000,000 light-years from us, is receding at a velocity half that of light.

The reason for Hubble's Law is taken to lie in the expansion of the universe itself — an expansion which can be made to follow from the equations set up by Einstein's

General Theory of Relativity (which, I hereby state firmly, I will *not* go into).

Given the expansion of the universe, now, how are Olbers' assumptions affected?

If, at a distance of 6,000,000,000 light-years a galaxy recedes at half the speed of light, then at a distance of 12,000,000,000 light-years a galaxy ought to be receding at the speed of light (if Hubble's Law holds). Surely, further distances are meaningless, for we cannot have velocities greater than that of light. Even if that were possible, no light, or any other 'message' could reach us from such a more-distant galaxy and it would not, in effect, be in our universe. Consequently, we can imagine the universe to be finite after all, with a 'Hubble radius' of some 12,000,000,000 light-years.[2]

But that doesn't wipe out Olbers' paradox. Under the requirements of Einsten's theories, as galaxies move faster and faster relative to an observer, they become shorter and shorter in the line of travel and take up less and less space, so that there is room for larger and larger numbers of galaxies. In fact, even in a finite universe, with a radius of 12,000,000,000 light-years, there might still be an infinite number of galaxies; almost all of them (paper-thin) existing in the outermost few miles of the Universe-sphere.

So Assumption 2 stands even if Assumption 1 does not; and Assumption 2, by itself, can be enough to ensure a star-bright sky.

But what about the red shift?

Astronomers measure the red shift by the change in position of the spectral lines, but those lines move only because the entire spectrum moves. A shift to the red is a shift in the direction of lesser energy. A receding galaxy delivers less radiant energy to the Earth than the same galaxy would deliver if it were standing still relative to us —

just because of the red shift. The faster a galaxy recedes the less radiant energy it delivers. A galaxy receding at the speed of light delivers no radiant energy at all no matter how bright it might be.

Thus, Assumption 3 fails! It would hold true if the universe were static, but not if it is expanding. Each succeeding shell in an expanding universe delivers less light than the one within because its content of galaxies is successively farther from us; is subjected to a successively greater red shift; and falls short, more and more, of the expected radiant energy it might deliver.

And because Assumption 3 fails, we receive only a finite amount of energy from the universe and the night sky is black.

According to the most popular models of the universe, this expansion will always continue. It may continue without the production of new galaxies so that, eventually, billions of years hence, our Galaxy (plus a few of its neighbours, which together make up the 'local cluster' of galaxies) will seem alone in the universe. All the other galaxies will have receded too far to detect. Or new galaxies may continuously form so that the universe will always seem full of galaxies, despite its expansion. Either way, however, expansion will continue and the night sky will remain black.

There is another suggestion, however, that the universe oscillates; that the expansion will gradually slow down until the universe comes to a moment of static pause, then begins to contract again, faster and faster, till it tightens at last into a small sphere that explodes and brings about a new expansion.

If so, then as the expansion slows the dimming effect of the red shift will diminish and the night sky will slowly brighten. By the time the universe is static the sky will be uniformly star-bright as Olbers' paradox required. Then, once the universe starts contracting, there will be a 'violet-shift' and the energy delivered will increase so that the sky will become far brighter and still brighter.

This will be true not only for the Earth (if it still existed in the far future of a contracting universe) but for any body of any sort in the universe. In a static or, worse still, a contracting universe there could, by Olbers' paradox, be no cold bodies, no solid bodies. There would be uniform high temperatures everywhere — in the millions of degrees, I suspect — and life simply could not exist.

So I get back to my earlier statement. The reason there is life on Earth, or anywhere in the universe, is simply that the distant galaxies are moving away from us.

In fact, now that we know the ins and outs of Olbers' paradox, might we, do you suppose, be able to work out the recession of the distant galaxies as a necessary consequence of the blackness of the night sky? Maybe we could amend the famous statement of the French philosopher René Descartes.

He said, 'I think, therefore I am!'

And we could add: 'I am, therefore the universe expands!'

CHAPTER SEVENTEEN

A GALAXY AT A TIME

Four or five years ago there was a small fire at a school two blocks from my house. It wasn't much of a fire, really, producing smoke and damaging some rooms in the basement, but nothing more. What's more, it was outside school hours so that no lives were in danger.

Nevertheless, as soon as the first piece of fire apparatus was on the scene the audience had begun to gather. Every idiot in town and half the idiots from the various contiguous towns came racing down to see the fire. They came by auto and by oxcart, on bicycle and on foot. They came with girl friends on their arms, with aged parents on their shoulders, and with infants at the breast.

They parked all the streets solid for miles around and after the first fire engine had come on the scene nothing more could have been added to it except by helicopter.

Apparently this happens every time. At every disaster, big or small, the two-legged ghouls gather and line up shoulder to shoulder and chest to back. They do this, it seems, for two purposes: (*a*) to stare goggle-eyed and slack-jawed at destruction and misery, and (*b*) to prevent the approach of the proper authorities who are attempting to safeguard life and property.

Naturally, I wasn't one of those who rushed to see the fire and I felt very self-righteously noble about it. However (since we are all friends), I will confess that this is not necessarily because I am free of the destructive instinct. It's just that a messy little fire in a basement isn't *my* idea of destruction; or a good, roaring blaze at the munitions dump, either.

If a star were to blow up, *then* we might have something.

Come to think of it, my instinct for destruction must be well developed after all, or I wouldn't find myself so fascinated by the subject of supernovas, those colossal stellar explosions.

Yet in thinking of them, I have, it turns out, been a piker. Here I've been assuming for years that a supernova was the grandest spectacle the universe had to offer (provided you were standing several dozen light-years away) but, thanks to certain 1963 findings, it turns out that a supernova taken by itself is not much more than a two-inch firecracker.

This realisation arose out of radio astronomy. Since World War II, astronomers have been picking up microwave (very short radio wave) radiation from various parts of the sky, and have found that some of it comes from our own neighbourhood. The Sun itself is a radio source and so are Jupiter and Venus.

The radio sources of the Solar System, however, are virtually insignificant. We would never spot them if we weren't right here with them. To pick up radio waves across the vastness of stellar distances we need something better. For instance, one radio source from beyond the Solar System is the Crab Nebula. Even after its radio waves have been diluted by spreading out for five thousand light-years before reaching us, we can still pick up what remains and impinges upon our instruments. But then the Crab Nebula represents the remains of a supernova that blew itself to kingdom come — the first light of the explosion reaching the Earth about 900 years ago.

But a great number of radio sources lie outside our Galaxy

altogether and are millions and even billions of light-years distant. *Still* their radio wave emanations can be detected and so they must represent energy sources that shrink mere supernovas to virtually nothing.

For instance, one particularly strong source turned out, on investigation, to arise from a galaxy 200,000,000 light-years away. Once the large telescopes zeroed in on that galaxy it turned out to be distorted in shape. After closer study it became quite clear that it was not a galaxy at all, but *two* galaxies in the process of collision.

When two galaxies collide like that, there is little likelihood of actual collisions between stars (which are too small and too widely spaced). However, if the galaxies possess clouds of dust (and many galaxies, including our own, do), these clouds will collide and the turbulence of the collision will set up radio-wave emission, as does the turbulence (in order of decreasing intensity) of the gases of the Crab Nebula, of our Sun, of the atmosphere of Jupiter, and of the atmosphere of Venus.

But as more and more radio sources were detected and pinpointed, the number found among the far-distant galaxies seemed impossibly high. There might be occasional collisions among galaxies but it seemed most unlikely that there could be enough collisions to account for all those radio sources.

Was there any other possible explanation? What was needed was some cataclysm just as vast and intense as that represented by a pair of colliding galaxies, but one that involved a single galaxy. Once freed from the necessity of supposing collisions we can explain any number of radio sources.

But what can a single galaxy do alone, without the help of a sister galaxy?

Well, it can explode.

But how? A galaxy isn't really a single object. It is simply a loose aggregate of up to a couple of hundred billion stars. These stars can explode individually, but how can we have an explosion of a whole galaxy at a time?

To answer that, let's begin by understanding that a galaxy isn't really as loose an aggregation as we might tend to think. A galaxy like our own may stretch out 100,000 light-years in its longest diameter, but most of that consists of nothing more than a thin powdering of stars — thin enough to be ignored. We happen to live in this thinly starred outskirt of our own Galaxy so we accept that as the norm, but it isn't.

The nub of a galaxy is its nucleus, a dense packet of stars roughly spherical in shape and with a diameter of, say, 10,000 light-years. Its volume is then 525,000,000,000 cubic light-years, and if it contains 100,000,000,000 stars, that means there is 1 star per 5·25 cubic light-years.

With stars massed together like that, the average distance between stars in the galactic nucleus is 1·7 light-years — but that's the average over the entire volume. The density of star numbers in such a nucleus increases as one moves towards the centre, and I think it is entirely fair to expect that towards the centre of the nucleus, stars are not separated by more than half a light-year.

Even half a light-year is something like 3,000,000,000,000 miles or 400 times the extreme width of Pluto's orbit, so that the stars aren't actually *crowded*; they're not likely to be colliding with each other, and yet . . .

Now suppose that, somewhere in a galaxy, a supernova lets go.

What happens?

In most cases, nothing (except that one star is smashed to flinders). If the supernova were in a galactic suburb — in our own neighbourhood, for instance — the stars would be so thinly spread out that none of them would be near enough to pick up much in the way of radiation. The incredible quantities of energy poured out into space by such a supernova would simply spread and thin out and come to nothing.

In the centre of a galactic nucleus, the supernova is not quite as easy to dismiss. A good supernova at its height is releasing energy at nearly 10,000,000,000 times the rate of our Sun. An object five light-years away would pick up a tenth as much energy per second as the Earth picks up from the Sun. At half a light-year from the supernova it would pick up ten times as much energy per second as Earth picks up from the Sun.

This isn't good. If a supernova let go five light-years from us we would have a year of bad heat problems. If it were half a light-year away I suspect there would be little left of earthly life. However, don't worry. There is only one star-system within five light-years of us and it is not the kind that can go supernova.

But what about the effects on the stars themselves? If our Sun were in the neighbourhood of a supernova it would be subjected to a bath of energy and its own temperature would have to go up. After the supernova is done, the Sun would seek its own equilibrium again and be as good as before (though life on its planets may not be). However, in the process, it would have increased its fuel consumption in proportion to the fourth power of its absolute temperature. Even a small rise in temperature might lead to a surprisingly large consumption of fuel.

It is by fuel consumption that one measures a star's age. When the fuel supply shrinks low enough, the star expands into a red giant or explodes into a supernova. A distant supernova by warming the Sun slightly for a year might cause it to move a century, or ten centuries closer to such a crisis. Fortunately, our Sun has a long lifetime ahead of it (several billion years), and a few centuries or even a million years would mean little.

Some stars, however, cannot afford to age even slightly. They are already close to that state of fuel consumption which will lead to drastic changes, perhaps even supernova-hood. Let's call such stars, which are on the brink, pre-supernovas. How many of them would there be per galaxy?

It has been estimated that there are an average of 3 supernovas per century in the average galaxy. That means that in 33,000,000 years there are about a million supernovas in the average galaxy. Considering that a galactic life span may easily be a hundred billion years, any star that's only a few million years removed from supernova-hood may reasonably well be said to be on the brink.

If, out of the hundred billion stars in an average galactic nucleus, a million stars are on the brink, then 1 star out of 100,000 is a pre-supernova. This means that pre-supernovas within galactic nuclei are separated by average distances of 80 light-years. Towards the centre of the nucleus, the average distance of separation might be as low as 25 light-years.

But even at 25 light-years, the light from a supernova would be only $1/250$ that which the Earth receives from the Sun, and its effect would be trifling. And, as a matter of fact, we frequently see supernovas light up one galaxy or another and nothing happens. At least, the supernova slowly dies out and the galaxy is then as it was before.

However, if the average galaxy has 1 pre-supernova in every 100,000 stars, particular galaxies may be poorer than that in supernovas — or richer. An occasional galaxy may be particularly rich and 1 star out of every 1,000 may be a pre-supernova.

In such a galaxy, the nucleus would contain 100,000,000 pre-supernovas, separated by an average distance of 17 light-years. Towards the centre, the average separation might be no more than 5 light-years. If a supernova lights up a pre-supernova only 5 light-years away it will shorten its life significantly, and if that supernova had been a thousand years from explosion before, it might be only two months from explosion afterwards.

Then, when it lets go, a more distant pre-supernova which has had its lifetime shortened, but not so drastically, by the first, may have its lifetime shortened again by the second and closer supernova, and after a few months *it* blasts.

On and on like a bunch of tumbling dominoes this would go, until we end up with a galaxy in which not a single supernova lets bang, but several million perhaps, one after the other.

There is the galactic explosion. Surely such a tumbling of dominoes would be sufficient to give birth to a coruscation of radio waves that would still be easily detectable even after it had spread out for a billion light-years.

Is this just speculation? To begin with, it was, but in late 1963 some observational data made it appear to be more than that.

It involves a galaxy in Ursa Major which is called M82 because it is number 82 on a list of objects in the heavens prepared by the French astronomer Charles Messier about two hundred years ago.

Messier was a comet-hunter and was always looking through his telescope and thinking he had found a comet and turning handsprings and then finding out that he had been fooled by some foggy object which was always there and was *not* a comet.

Finally, he decided to map each of 101 annoying objects that were foggy but were not comets so that others would not be fooled as he was. It was that list of annoyances that made his name immortal.

The first on his list, M1, is the Crab Nebula. Over two dozen are globular clusters (spherical conglomerations of densely strewn stars), M13 being the Great Hercules Cluster, which is the largest known. Over thirty members of his list are galaxies, including the Andromeda Galaxy (M31) and the Whirlpool Galaxy (M51). Other famous objects on the list are the Orion Nebula (M42), the Ring Nebula (M57), and the Owl Nebula (M97).

Anyway, M82 is a galaxy about 10,000,000 light years from Earth which aroused interest when it proved to be a strong radio source. Astronomers turned the 200-inch telescope upon it and took pictures through filters that blocked all light except that coming from hydrogen ions. There was reason to suppose that any disturbances that might exist would show up most clearly among the hydrogen ions.

They did! A three-hour exposure revealed jets of hydrogen up to a thousand light-years long, bursting out of the galactic nucleus. The total mass of hydrogen being shot out was the equivalent of at least 5,000,000 average stars. From the rate at which the jets were travelling and the distance they had covered, the explosion must have taken place about 1,500,000 years before. (Of course, it takes light ten million years to reach us from M82, so that the explosion took place 11,500,000 years ago, Earth-time — just at the beginning of the Pleistocene Epoch.)

M82, then, is the case of an exploding galaxy. The energy expended is equivalent to that of five million supernovas formed in rapid succession, like uranium atoms undergoing fission in an atomic bomb — though on a vastly greater scale, to be sure. I feel quite certain that if there had been any life anywhere in that galactic nucleus, there isn't any now.

In fact, I suspect that even the outskirts of the galaxy may no longer be examples of prime real estate.

Which brings up a horrible thought ... Yes, you've guessed it!

What if the nucleus of our own dear Galaxy explodes? It very likely won't, of course (I don't want to cause fear and despondency among the Gentle Readers), for exploding galaxies are probably as uncommon among galaxies as exploding stars are among stars. Still, if it's not going to happen, it is all the more comfortable then, as an intellectual exercise, to wonder about the consequences of such an explosion.

To begin with, we are not in the nucleus of our Galaxy but far in the outskirts and in distance there is a modicum of safety. This is especially so since between ourselves and the nucleus are vast clouds of dust that will effectively screen off any visible fireworks.

Of course, the radio waves would come spewing out, through dust and all, and this would probably ruin radio astronomy for millions of years by blanking out everything else. Worse still would be the cosmic radiation that might rise high enough to become fatal to life. In other words, we might be caught in the fallout of that galactic explosion.

Suppose, though, we put cosmic radiation to one side, since the extent of its formation is uncertain and since consideration of its presence would be depressing to the spirits. Let's also abolish the dust clouds with a wave of the speculative hand.

Now we can see the nucleus. What does it look like without an explosion?

Considering the nucleus to be 10,000 light-years in diameter and 30,000 light-years away from us, it would be visible as a roughly spherical area about 20° in diameter. When entirely above the horizon it would make up a patch about $1/_{65}$ of the visible sky.

Its total light would be about 30 times that given off by

Venus at its brightest, but spread out over so large an area it would look comparatively dim. An area of the nucleus equal in size to the full Moon would have an average brightness only 1/200,000 of the full Moon.

It would be visible then as a patch of luminosity broadening out of the Milky Way in the constellation of Sagittarius, distinctly brighter than the Milky Way itself; brightest at the centre, in fact, and fading off with distance from the centre.

But what if the nucleus of our Galaxy exploded? The explosion would take place, I feel certain, in the centre of the nucleus, where the stars were thickest and the effect of one pre-supernova on its neighbours would be most marked. Let us suppose that 5,000,000 supernovas are formed, as in M82.

If the nucleus has pre-supernovas separated by 5 light-years in its central regions (as estimated earlier in the chapter, for galaxies capable of explosion), then 5,000,000 pre-supernovas would fit into a sphere about 850 light-years in diameter. At a distance of 30,000 light-years, such a sphere would appear to have a diameter of 1·6°, which is a little more than three times the apparent diameter of the full Moon. We would therefore have an excellent view.

Once the explosion started, supernova ought to follow supernova at an accelerating rate. It would be a chain reaction.

If we were to look back on that vast explosion millions of years later, we could say (and be roughly correct) that the centre of the nucleus had all exploded at once. But this is only roughly correct. If we actually watch the explosion in process, we will find it will take considerable time, thanks entirely to the fact that light takes considerable time to travel from one star to another.

When a supernova explodes, it can't affect a neighbouring

pre-supernova (5 light-years away, remember) until the radiation of the first star reaches the second — and that would take 5 years. If the second star was on the far side of the first (with respect to ourselves), an additional 5 years would be lost while the light travelled back to the vicinity of the first. We would therefore see the second supernova 10 years later than the first.

Since a supernova will not remain visible to the naked eye for more than a year or so even under the best conditions (at the distance of the Galactic nucleus), the second supernova would not be visible until long after the first had faded off to invisibility.

In short, the 5,000,000 supernovas, forming in a sphere 850 light-years in diameter, would be seen by us to appear over a stretch of time equal to roughly a thousand years. If the explosions started at the near edge of that sphere so that radiation had to travel away from us and return to set off other supernovas, the spread might easily be 1,500 years. If it started at the far end and additional explosions took place as the light of the original explosion passed the pre-supernovas *en route* to ourselves, the time-spread might be considerably less.

On the whole, the chances are that the Galactic nucleus would begin to show individual twinkles. At first there might be only three or four twinkles a decade, but then, as the decades and centuries passed, there would be more and more until finally there might be several hundred visible at one time. And finally, they would all go out and leave behind dimly glowing gaseous turbulence.

How bright will the individual twinkles be? A single supernova can reach a maximum absolute magnitude of -17. That means if it were at a distance of 10 parsecs (32·5 light-years) from ourselves, it would have an apparent magnitude of —17, which is 1/10,000 the brightness of the Sun.

At a distance of 30,000 light-years, the apparent magnitude of such a supernova would decline by 15 magnitudes. The apparent magnitude would now be —2, which is about the brightness of Jupiter at its brightest.

This is quite a startling statistic. At the distance of the nucleus, no ordinary star can be individually seen with the naked eye. The hundred billion stars of the nucleus just make up a luminous but featureless haze under ordinary conditions. For a single star, at that distance, to fire up to the apparent brightness of Jupiter is simply colossal. Such a supernova, in fact, burns with a tenth the light intensity of an entire non-exploding galaxy such as ours.

Yet it is unlikely that every supernova forming will be a supernova of maximum brilliance. Let's be conservative and suppose that the supernovas will be, on the average, two magnitudes below the maximum. Each will then have a magnitude of 0, about that of the star Arcturus.

Even so, the 'twinkles' would be prominent indeed. If humanity were exposed to such a sight in the early stages of civilisation, they would never make the mistake of thinking that the heavens were eternally fixed and unchangeable. Perhaps the absence of that particular misconception (which, in actual fact, mankind laboured under until early modern times) might have accelerated the development of astronomy.

However, we can't see the Galactic nucleus and that's that. Is there anything even faintly approaching such a multi-explosion that we *can* see?

There's one conceivable possibility. Here and there, in our Galaxy, are to be found globular clusters. It is estimated there are about 200 of these per galaxy. (About a hundred of our own clusters have been observed, and the other hundred are probably obscured by the dust clouds.)

These globular clusters are like detached bits of galactic nuclei, 100 light-years or so in diameter and containing from 100,000 to 10,000,000 stars — symmetrically scattered about the galactic centre.

The largest known globular cluster is the Great Hercules Cluster, M13, but it is not the closest. The nearest globular cluster is Omega Centauri, which is 22,000 light-years from us and is clearly visible to the naked eye as an object of the fifth magnitude. It is only a point of light to the naked eye, however, for at that distance even a diameter of 100 light-years covers an area of only about 1·5 minutes of arc in diameter.

Now let us say that Omega Centauri contained 10,000 pre-supernovas and that every one of these exploded at their earliest opportunity. There would be fewer twinkles altogether, but they would appear over a shorter time interval and would be, individually, twice as bright.

It would be a perfectly ideal explosion, for it would be unobscured by dust clouds; it would be small enough to be quite safe; and large enough to be sufficiently spectacular for anyone.

And yet, now that I've worked up my sense of excitement over the spectacle, I must admit that the chances of viewing an explosion in Omega Centauri are just about nil. And even if it happened, Omega Centauri is not visible in New England and I would have to travel quite a bit southward if I expected to see it high in the sky in full glory — and I don't like to travel.

Hmm . . . Oh well, anyone for a neighbourhood fire?

FOOTNOTES FOR CHAPTERS

INTRODUCTION
1 Don't tell me you don't have one!

CHAPTER 1
1 This has been published: *The Moon* (Follett, 1967).

CHAPTER 2
1 This was *Lucky Star and the Ocean of Venus* (Doubleday, 1954). Since this article first appeared it has appeared as a paperback under my own name, with a preliminary note explaining the outdated astronomy.
2 Mars probes, since this article was written, show conditions on Mars to be even harsher than had been thought — colder, for instance. They also showed a very variegated surface, with volcanic areas and possible erosion effects. The question of life on Mars is still rather an open question. There are even suggestions based on recent data that Mars enjoys periodic eras of mild conditions with both air and water in considerable quantities.
3 Since this article was written, geologists have grown convinced that, through continental drift, the Atlantic Ocean has been formed quite recently (as geologic history goes). Its borders fit together quite closely and could not possibly have been hammered out by large meteor strikes. I doubt that Dr. Dachille's enthusiasm has gained many adherents, but that doesn't mean that meteorite strikes haven't taken place at various sites and times.
4 No meteoric matter has been located on the site so far. Could it have been a collision with a small comet made up of substances that vaporize after impact? Could it have been a small piece of anti-matter which explodes on contact with ordinary matter and leaves nothing behind? Alas, no one knows.

5 Two years after this article first appeared, I repeated this conjecture in *Asimov's Guide to the Bible, Volume 1, The Old Testament* (Doubleday, 1968). So far, no archaeologists have written to say what an inspired guess this is. It's a shame the way we geniuses go unrecognized.

6 I was wrong here. Since this article was written, archaeologists have discovered that a small island in the southern Aegean exploded in a gigantic volcanic explosion that set up a tsunami that destroyed Cretan civilisation. It was this exploding island in about 1400 B.C. that left vague memories that turned up as Plato's Atlantis a thousand years later, No meteorite, but anyway a catastrophe.

7 In 1964, two years before this article appeared, another Earth-grazer was discovered and was named Toro. It never approaches closer to us than 9 million miles, but its orbit is locked into that of Earth's, so that it never recedes past a certain limit. It circles the Sun 5 times in 8 years in a complex dance about the Earth. It is a kind of 'quasi-moon' and seems to present Earth with no danger.

CHAPTER 3

1 Professor McLaughlin, who has died, alas, since this article first appeared, is the father of a well-known science-fiction fan of the same name.

2 Since this was written, the effect of Venus on artificial probes passing near it has given astronomers the chance to calculate its mass with considerable precision.

CHAPTER 4

1 Oh, well, why be coy. My recently published book *The Early Asimov* (Doubleday, 1972) tells all my secrets anyway. My first published story was 'Marooned Off Vesta' and it appeared in the March 1939 *Amazing Stories*.

2 In 1973, *Pioneer 10* passed through the asteroid zone on the way to Jupiter. It had no trouble and found fewer particles than had been expected.

3 In late 1967, some four years after this article first appeared, another satellite of Saturn was discovered which was even closer to the planet than Mimas. This new satellite, Janus, has a period of revolution of 18 hours. Its existence does not affect the line of argument here.

4 There is of course the 'quasi-moon' Toro, mentioned in chapter 2, footnote 7.

CHAPTER 5

1 The atmosphere is not necessarily gaseous. It is made up of compounds which would be gases under earthly conditions, but which under Jupiter's temperatures and pressures might be liquid or even solid at certain depths.

2 As a matter of fact, a picture currently popular is of a Jupiter made up almost entirely of hydrogen and helium (with fourteen atoms of the former for every one of the latter). Under great pressures, hydrogen solidifies and gains metallic properties, so that Jupiter's core is of 'metallic hydrogen'. Actually, though all speculations concerning Jupiter's inner structure remain, as yet, only intelligent guesses.

3 I know that the atmosphere is thicker than eight miles, and that in fact it has no fixed thickness. However, I am taking the earth's atmosphere — and shall later calculate its volume — only to the top of its cloud layers, which is what we do for the giant planets.

4 Of course, if the hydrogen-helium picture of Jupiter is correct, this whole argument is seriously compromised.

CHAPTER 6

1 And not just trying. As we all know, American astronauts first reached the Moon in 1969, some years after this article was first published and have revisited it several times since then.

2 Rocks brought back from the Moon since 1969 show no traces of having ever had any water content, which is disheartening for the prospects of Moon colonisation.

3 Actually, Pluto is now thought to be only the size of Mars, rather than the size of Earth, as had been assumed when this article first appeared.

4 In 1971 a Mars-probe photographed Phobos, the larger satellite, and showed it to be potato-like in appearance (honest!) and 16 miles in its longest diameter.

5 It was called 'Strikebreaker' and you'll find it in my collection *Nightfall and Other Stories* (Doubleday, 1969).

CHAPTER 7

1 Heavens! Since this article first appeared in January 1964, astronomers have found that both Mercury and Venus rotate with respect to the Sun after all. Mercury's period of rotation is 59 days and that of Venus is 243 days.

2 The length of the sidereal day for Venus and Mercury is given here correctly, in line with the new knowledge concerning their rotations. The figures given in the article when it first appeared turned out, of course to be wrong.

CHAPTER 8

1 In 1972, actually, calculations have been made from deviations in the orbit of Halley's comet that aren't accounted for by the known planets. These calculations show that a Tenth Planet may indeed exist with properties rather like those described in this article, which was written in early 1960. The Tenth Planet, however, hasn't been seen and later calculations make its existence — at least on the Halley's comet basis — unlikely.

CHAPTER 9

1 The satellite Janus was discovered four years after this article was first published, so it wasn't included. I am adding it now.
2 I am including Janus in the list, remember, although it was unknown when the article first appeared.
3 The Roche limit only holds true exactly for satellites of more than a certain size and a few other qualifications but we don't have to worry about that here.
4 This article was written in 1963. It was hoped that once the Moon was actually reached, a study of its surface might tell us how it originated and whether it was captured or not. So far, however, although the Moon has been visited several times, no answer has been obtained. The information we have received offers more puzzles than answers.

CHAPTER 10

1 This article first appeared in 1960 and shows me with my usual optimism. The Moon has been 'conquered', to be sure, but how far we will be able to continue space exploration in the present mood of disenchantment, I can't say.
2 When this article first appeared I said it was 'the radiation pressure of sunlight' that formed the comet's tail. That's what astronomers had thought but radiation pressure just wasn't strong enough. The existence of the Solar Wind (particles driven away from the Sun in

every direction) was made clear in 1958, two years before this article was written, but its efficacy in connection with comet's tails wasn't straightened out in time for me to include it.

CHAPTER 11

1 I wrote such a story myself in 1954. It was called 'Sucker Bait' and you will find it in my book *The Martian Way and Other Stories* (Doubleday, 1955). At least, I tried to rationalise the situation in that story, but in 1941, I wrote 'Nightfall' about a planet with *six* suns. It is collected in *Nightfall and Other Stories* (Doubleday, 1969). Despite the fact that the astronomic situation in 'Nightfall' is enormously unlikely, it remains my most popular short story.

CHAPTER 12

1 There are, of course, 'neutron stars' which are far smaller and far denser than any white dwarf can be. They are not mentioned in this article for the very good reason that the article first appeared in 1963, and neutron stars were not discovered until 1968. However, the omission of neutron stars does not invalidate any of the information in this article.

2 Actually, it has recently been reported that Jupiter radiates somewhat more heat than can be accounted for by solar radiation. Maybe it has nuclear reactions going on in its compressed centre and it is a very small and very un-warm star.

CHAPTER 13

1 I have always thought, deep in my heart, that this notion of mapping the heavens on the globe of the earth was the most brilliant I have ever had. In the dozen years since this article first appeared in 1961, I never even had a single letter saying, 'Gee, Isaac, that was brilliant.' Oh, well . . .

CHAPTER 15

1 And nine years after this first appeared, the Moon was actually reached — in 1969.

2 Or, more likely, we *now* know, to a neutron star.

3 It turned out to be better than that. New studies ten years after this article first appeared showed it to be a neutron star.

4 If they ended up as neutron stars, very unlikely in the case of ordinary novae, they might not be seen optically, but they might be detected as 'pulsars', bodies producing very rapid pulses of radio beams.

CHAPTER 16

1 This article was first published in 1964. Robyn is now at college age and we are both still very fond of 'Peanuts', whose characters are still in the fourth grade — or less.

2 In 1973, a quasar was detected at that distance and there were immediate newspaper headlines to the effect we had seen the 'end of the universe'.

INDEX